VISUALIZING STATISTICAL MODELS AND CONCEPTS

Additional Volumes in Preparation

VISUALIZING STATISTICAL MODELS AND CONCEPTS

R. W. FAREBROTHER

Faculty of Economics and Social Studies
Victoria University of Manchester
Manchester, England

Illustrated by Michaël Schyns

MARCEL

DEKKER

MARCEL DEKKER, INC.   NEW YORK · BASEL

ISBN: 0-8247-0718-4

This book is printed on acid-free paper.

Headquarters
Marcel Dekker, Inc.
270 Madison Avenue, New York, NY 10016
tel: 212-696-9000; fax: 212-685-4540

Eastern Hemisphere Distribution
Marcel Dekker AG
Hutgasse 4, Postfach 812, CH-4001 Basel, Switzerland
tel: 41-61-261-8482; fax: 41-61-261-8896

World Wide Web
http://www.dekker.com

The publisher offers discounts on this book when ordered in bulk quantities. For more information, write to Special Sales/Professional Marketing at the headquarters address above.

Current printing (last digit):
10 9 8 7 6 5 4 3 2 1

PRINTED IN THE UNITED STATES OF AMERICA

Preface

Until relatively recent times, geometrical and mechanical analogies played an essential role in the development of mathematical models for the physical sciences. The same is true of the models and concepts of mathematical statistics. However, from the beginning of the twentieth century, there has been a growing tendency for statisticians to turn away from such matters, so much so that typical modern statisticians have few, if any, geometrical insights into the nature of the fitting problems they are dealing with, and no mechanical insights whatsoever.

This almost total absence of geometrical and mechanical analogies from modern mathematical statistics constitutes a sad depletion of the subject. For, although visual representations of this type are no longer required as the basis of suitable analogue computing devices, they are still valuable as a means of enhancing the practitioner's understanding of the concepts involved. In this book, I attempt to remedy this parlous situation by suggesting suitable geometrical and mechanical analogies for some of the more accessible statistical techniques. In particular, I address the problem of determining the position of a multivariate location parameter and the problem of fitting a plane or curved surface to multivariate data when the goodness of fit criterion is variously defined by the sum of the squared deviations, the sum of the absolute deviations, the median absolute deviation, or the maximum absolute deviation.

This book is intended for use as auxiliary reading by students of mathematical statistics and as a reference source for research workers in this area. It may also serve as the basis of a (largely heuristic) introductory course in statistics for students of engineering, physics, and technology who are sufficiently familiar with the mechanical concepts underlying this treatment of the subject.

To enhance its role as a reference work, the successive chapters of the book have been made as independent of one another as possible. However, I have reluctantly decided not to investigate any of the linear or nonlinear programming procedures in detail, as to do so would presume too much of readers' knowledge of this area and would overburden the text with numerical examples.

Hopefully, figures similar to those presented in the text will soon make their appearance in commercially produced packages of statistical graphics.

The successive drafts of this book were typed by my wife, Sheila, and myself. The present version of the book was prepared from our final typescript by Keyword Typesetting Services under the direction of Marcel Dekker, Inc. I am indebted to Stephen Pollock of Queen Mary and Westfield College, University of London, England, for translating the original draft into *Plain TEX*; to Götz Trenkler and Sven-Oliver Troschke of the University of Dortmund, Germany, for their detailed criticism of a later version; but, above all, I am indebted to Michaël Schyns of the University of Namur (FUNDP), Belgium, for creating the figures given in the text, and to my wife, Sheila, for all her help in the preparation of this book, not least for her discovery of a person skilled in statistical graphics amongst those attending the Third L_1-norm Conference in Neuchâtel, Switzerland, in August 1997. Without their aid this project would surely have foundered.

R. W. Farebrother

Contents

CHAPTER 2

Abstract Geometrical and Mechanical Representations   13

CHAPTER 3

Mechanical Models for Multidimensional Medians   29

CHAPTER 4

Method of Least Squared Deviations   63

CHAPTER 5

Method of Least Absolute Deviation   97

CHAPTER 6

Minimax Absolute Deviation Method   143

CHAPTER 11

Multivariate Generalisations of the Method of Least Squares   219

VISUALIZING STATISTICAL MODELS AND CONCEPTS

CHAPTER 1

Introduction

1.1　Introductory Remarks

Geometrical diagrams are a familiar feature of elementary statistical textbooks. Amongst the most familiar of such diagrams are the standard (side-by-side) and the partitioned (end-to-end) bar charts or histograms of Figures 1.1 and 1.2 and the (observation-space) representations of the conventional and orthogonal linear regression models of Figures 1.3 and 1.4.

But these familiar representations of statistical procedures are entirely static and can only benefit from the admixture of a certain amount of mechanical interpretation. However, such interpretations need to be chosen with care if the resulting combination is to offer the clearest possible insight into the essential nature of the relevant statistical procedure.

The psychological importance of clear visual impressions of the mathematical concepts underlying particular theories has been stressed by Friendly (1995), Sall (1991a, 1991b), and Skemp (1971). In this book, we shall find that many of the important concepts of mathematical statistics can be associated with physical models; and that the optimality criteria of statistical estimation procedures can often be interpreted in terms of the concept of potential energy. For example, in Chapters 4 and 5 we shall find that our intuitive understanding of the linear regression model of Figures 1.3 and 1.4 is significantly enhanced if we interpret the line segments joining

1

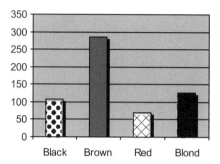

Figure 1.1   Side-by-side bar chart.

the observed points to the fitted line as if they were springs or strings under tension connected to a rigid rod. Similarly, in Chapter 9, we shall find that our understanding of the dynamic model underlying the bar charts or histograms of Figures 1.1 and 1.2 is much improved if we interpret the rectangular areas in these figures as if they were the cross-sections of adjacent containers holding appropriate quantities of liquid or gas.

    In this context, we should also mention that there is a class of complex mathematical problems that are difficult to solve without the insight gained from appropriate mechanical models. The problems discussed in this book are not of this type. Nevertheless, at the end of Chapter 3, we shall take the opportunity of illustrating this point by giving a brief outline of Richard Courant's (1940) use of

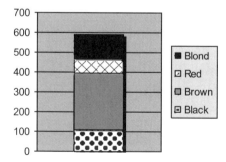

Figure 1.2   End-on-end bar chart.

physical analogies in solving a class of difficult problems in the calculus of variations.

1.2 The Role of Geometrical Models in Statistics

Examining the simple linear regression problem in greater detail, we find that we are given a set of n observations on two statistical variables X and Y. From the perspective of analytical geometry, it is natural to represent these n pairs of observations by a set of n points in a two-dimensional Cartesian plane defined by a pair of orthogonal axes marked with appropriate ranges of values of X and Y. Similarly, the hypothetical linear relationship underlying the observations on these two variables can be represented by a straight line in the same Cartesian plane. For any straight line in this plane, we may measure the distances from the points to the line, and we may define a goodness of fit criterion which is an increasing function of the absolute values of these distances.

Our problem is to identify a line that minimises the value of the selected optimality criterion. In particular, if the optimality criterion is the sum of the squared deviations of the points from the line, and if these deviations are measured perpendicular to the x-axis, then the best fitting line is the usual least squares line, see Figure 1.3.

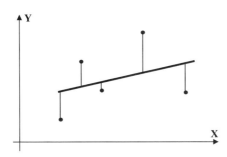

Figure 1.3 Conventional line fitting problem.

On the other hand, if distances are measured perpendicular to the fitted line, then the best fitting line is the orthogonal least squares line, see Figure 1.4.

A second application of geometry to the regression problem arises when the elliptical contours of the bivariate normal distribution of X and Y is directly observed, or we have sufficiently many observations to reconstruct these elliptical contours. In this context, the orthogonal regression line described above corresponds to the major axis of the elliptical contours, see Figure 1.5.

To generate the usual y on x regression line, we begin by finding points of tangency between the ellipses and lines drawn parallel to the y-axis. We note that these points of tangency lie on a straight line which is known as the y on x regression line, see Figure 1.6.

The alternative x on y regression line can be generated in a similar way by considering the points of tangency to a set of lines drawn parallel to the x-axis, see Figure 1.7.

In elementary statistical textbooks, the directions indicated by the x- and y-axes are often referred to as the horizontal and vertical directions respectively. However, it is not convenient to use this terminology in the present book as we shall be concerned with models that involve the representation of the optimality function as a three-dimensional geometrical figure lying vertically above the (horizontal) plane defined by these two axes. For this reason, the vertical direction will be strictly reserved for the representation of deviations from the fitted relationship and closely related func-

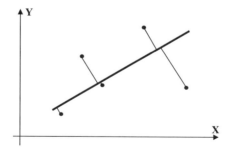

Figure 1.4   Orthogonal line fitting problem.

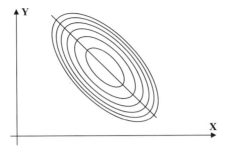

Figure 1.5 Orthogonal regression problem.

tions. The y-axis will usually be said to run in a south–north direction in the horizontal plane and the x-axis in a west–east direction. Thus, the usual nomenclature will only be employed when a function of the observed deviations is plotted against a single explanatory variable, as occurs in the opening subsections of Chapters 3 and 4.

1.3 The Analogy Implicit in Some Statistical Nomenclature

Having briefly examined the role of several geometrical concepts in statistical analyses, we can proceed to outline four statistical concepts that originate in geometry and mechanics.

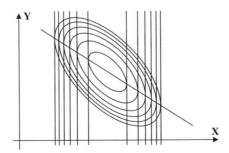

Figure 1.6 Conventional y on x regression problem.

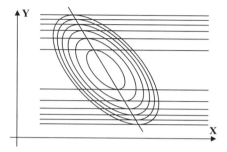

Figure 1.7 Conventional x on y regression problem.

Density: The concept of the density of a statistical distribution is readily interpreted in geometrical terms. Suppose that we wish to compare the density of the populations in two or more towns. We may begin by constructing a set of plane figures whose areas are proportional to the land areas of the corresponding towns. These plane figures may be scale representations of the geographical shape of the towns or they may be stylised representations in the form of discs or squares. The shape of the plane figures is immaterial. What is essential is that the area of the figures should be proportional to the areas of the associated towns.

Inside each figure we place a set of points whose numbers are proportional to the size of the population of the corresponding towns. The points should be of equal size and as evenly spaced as possible. The density of the points will show the densities of the populations of the corresponding towns. The more densely packed the points are in the available areas, the more densely packed will the corresponding populations be in reality.

Thus, a visual inspection of Figure 1.8 immediately reveals that the town represented by the rectangle on the right covers a greater area than does the town represented by the rectangle on the left. Further, the population of the town on the right is more crowded together than that of the town on the left, as the points are more densely packed into the rectangle on the right. In addition, for the particular instance illustrated here, it is possible to deduce that the population of the town on the right is greater than that on the left. On the other hand, if the town represented by the

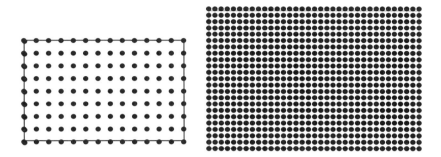

Figure 1.8 Relative densities of two populations.

smaller rectangle had had the higher density, then it would have been necessary to count the points in the rectangles before the populations of the two towns could be compared.

Weight: The idea of the weight of an observation is quite old. It features in the early work on the method of least squares of Carl Friedrich Gauss and Pierre Simon Laplace dating from the beginning of the nineteenth century. The idea was expounded in greater detail by George Boole in his account of the language of chance in 1857. [We shall not give specific references to these and later historical remarks. Interested readers are referred to Farebrother (1999), Hald (1998), and Stigler (1986) for details.]

As we might expect, an observation with greater weight has more influence on the estimated values of the parameters of a model than does one of lesser weight. The term weight also denotes the level of certainty, and the more weight attached to the estimate of a parameter value, the less will this value be affected by subsequent observations.

Laplace also employed the concept of weight in his analysis of the problem of minimising the sum of the absolute deviations. We shall tend to follow mechanical usage and reserve the word *weight* for the problem of least absolute deviations, see Chapter 5. We shall use the alternative word *modulus* in the context of the least squares problem of Chapter 4. However, we shall not illustrate this concept here. Instead, we refer readers to the next section and to Chapters 3 and 5 for the relevant diagrams.

Moment: If we are given a set of observations, represented as points distributed along a line, together with an arbitrary point on the same line, then we may rotate the given points about an axis passing through the arbitrary point. The more disperse is the distribution of the points, the more reluctant will the system be to rotate about the chosen axis under the influence of a given rotational force. The degree of inertia of the system is quantified by the sum of the squares of the distances of the points from the axis of rotation. This sum is known as the second moment or moment of inertia of the distribution. Moments of other orders are defined in a similar way by summing the relevant power of the distances from the given points to the axis.

Degrees of Freedom: The number of degrees of freedom of a particular problem records the number of entities that may be varied independently of each other. Like the concepts of *weight* and *moment*, this concept has been adopted from physics.

1.4   A Simple Mechanical Model for the Arithmetic Mean

As a representation of the arithmetic mean, we consider a chemical balance in a state of equilibrium. Suppose that we are given a set of n observations on a single variable Y. Then we may represent these observations as a scatter of points on the horizontal y-axis, and we may treat this axis as a horizontal beam with unit weights attached at the given points. Our problem is to place a fulcrum at the point of balance of the weighted beam.

Wherever we place the fulcrum, the weights to the right of that point will cause the beam to rotate in a clockwise direction, whilst weights to the left will cause it to rotate in an anticlockwise direction. The turning effect of a weight is measured by the product of the weight with its horizontal distance from the fulcrum, and is known as a *couple*.

Given the position of the fulcrum, we may determine the net clockwise couple of the system as a whole by summing the couples associated with each of the n weights. If this sum is positive then the beam will rotate in a clockwise direction, and if negative then it will rotate in an anticlockwise direction. Clearly, the system will be in equilibrium when the fulcrum is placed at the point which sets this sum equal to zero. The point of balance of the weighted beam corresponds to the arithmetic mean of the observations.

Figures 1.9a and 1.9b illustrate the case of a beam loaded with two unit weights. In the situation illustrated in Figure 1.9a, the fulcrum has been placed near the weight on the left and the beam will tend to rotate in a clockwise direction about this point. By contrast, in the situation illustrated in Figure 1.9b, the fulcrum has been placed exactly halfway between the two weights and the beam will remain in its present position.

This mechanical model may be generalised to yield valuable representations of weighted or unweighted multivariate means, see Chapter 10. However, it is unsatisfactory as a general analogy since it cannot be applied to more advanced statistical techniques without establishing a dynamic concept of which the net clockwise couple is the derivative. Even then, it is not easy to accommodate fitting procedures that are more general than the method of orthogonal least squares. See Chapter 4 for our preferred quasi-static model of the arithmetic mean.

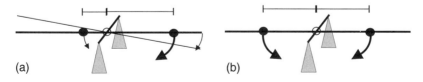

(a)   (b)

Figure 1.9 A loaded beam (a) when not in equilibrium, and (b) when in equilibrium.

1.5 Mechanical Models in Science

Like the chemical balance of Section 1.4, almost all of the physical models discussed in later chapters are drawn from classical mechanics. However, we are not restricted to analogies from this source as several of the physical sciences exhibit similar features. Indeed, Franksen (1969) has identified a group of quasi-static models from classical mechanics, electrical network theory, and thermodynamics (which he renames thermostatics), whose basic concepts may be employed as analogies for the corresponding concepts of other sciences in this group, and also for the corresponding models of economics and statistics.

Although, in this book, we shall concentrate on analogies drawn from classical mechanics, we shall mention models drawn from hydrostatics in Chapter 9. In principle, we may extend our scope to include models based on the flow of liquids in systems of pipes or on the flow of currents in electrical networks. But, such preliminary analyses as we have been able to conduct with hydrodynamic and electrical network theory suggest that these subject areas cannot offer significant insights into the workings of statistical models. Models of this type are, however, employed in economics where they are associated with the names of François Quesnay and Alban William Phillips. Similarly, we may base our models on systems of particles in motion under the effect of mutual gravitation; or, indeed, on collections of molecules in equilibrium as crystalline structures under a balance of forces. But, any attempt to address these models would take the analysis beyond the competence of the author, if not that of the reader.

Thus, although it is possible to base statistical models on a wide class of paradigms suggested by a variety of physical processes, we shall concentrate on those approaches that seem most accessible and potentially most valuable to students of statistics.

1.6 Possible Uses of This Book

At first sight, it might seem that this book is specifically addressed to students and research workers who have followed a conventional course in mathematical statistics with little or no support from geometrical or mechanical models. And, viewed in this light, it will serve as the basis of a supplementary course at the graduate or advanced undergraduate level for conventionally trained statisticians.

However, a more detailed examination of the contents will establish that there are two sub-themes within the unifying framework offered by the concept of potential energy, namely those associated with the use or non-use of the differential calculus and projective geometry duality. If the latter aspect of the subject is eliminated, then the remaining material in Section 2.1, and in Chapters 1, 3, 4, 9, 10, and 11 could be used as the basis of an elementary introduction to statistical estimation procedures for engineers, physicists, technologists, and others with a solid grounding in classical mechanics.

Alternatively, if we omit any mention of the differential calculus, then the remaining material in Chapters 1, 2, 3, 5, 6, 7, and 8 can be presented in the form of a geomechanical supplement to a conventional discussion of the line fitting problem for students of linear and nonlinear programming.

In addition, some of the mechanical models described in this book may be suitable for use by professional statisticians in their consultancy work when trying to explain some of their more abstruse estimation problems to clients with little knowledge of statistics. For instance, the contrast between the results generated by the spring models of Chapter 4 and the string models of Chapter 5 may help to justify the use of a more robust fitting procedure than the method of least squares. Similarly, the size of the strain in the springs of the models of Chapter 4 or the tension in the strings of the models of Chapter 5 may help to demonstrate the unreasonableness of some of the heuristic constraints championed by some clients.

References

Courant, R. (1940), Soap film experiments with minimal surfaces, *American Mathematical Monthly* **47**: 167–174.

Farebrother, R. W. (1987), Mechanical representations of the L_1 and L_2 estimation problems, in Y. Dodge (Ed.), *Statistical Data Analysis Based on the L_1-Norm and Related Methods*, North-Holland Publishing Company, Amsterdam, 455–464.

Farebrother, R. W. (1999), *Fitting Linear Relationships: A History of the Calculus of Observations 1750–1900*, Springer-Verlag, New York.

Franksen, O. I. (1969), Mathematical programming in economics by physical analogies, *Simulation* June, 297–314; July, 25–42; August, 63–87.

Friendly, M. (1995), Conceptual and visual models for categorical data, *The American Statistician* **49**: 153–160.

Hald, A. (1998), *A History of Mathematical Statistics from 1750 to 1930*, John Wiley and Sons, New York.

Sall, J. (1991a), The conceptual model behind the picture, *ASA Statistical Computing and Statistical Graphics Newsletter* **2**: 5–8.

Sall, J. (1991b), The conceptual model for categorical responses, *ASA Statistical Computing and Statistical Graphics Newsletter* **3**: 33–36.

Skemp, R. R. (1971), *The Psychology of Learning Mathematics*, Penguin Books, Harmondsworth, England.

Stigler, S. M. (1986), *The History of Statistics: The Measurement of Uncertainty Before 1900*, Harvard University Press, Cambridge, Massachusetts.

CHAPTER 2

Abstract Geometrical and Mechanical Representations

2.1 Bayesian and non-Bayesian Likelihood Functions

2.1.1 Likelihood Functions as Abstract Potential Energy Functions

In this chapter, we initiate our discussion of a wide range of disparate statistical, geometrical, and mechanical concepts which will prove useful to a greater or lesser degree in subsequent chapters. The first of these concepts is the (Bayesian or non-Bayesian) likelihood function and the associated hypothesis test.

Suppose that X is a random variable with a continuous probability density function $f(x; \theta)$ characterised by the value of a single unknown parameter θ. Then the probability density associated with a particular value $X = x_i$ of this variable is given by the corresponding value of the function

$$f(xi; \theta)$$

and the joint probability density of a set of n observations $x_1, x_2, \ldots, x_n$ by the likelihood function

$$L(\theta; x_1, x_2, \ldots, x_n) = \prod_i f(x_i; \theta)$$

In certain circumstances, this function of $x_1, x_2, \ldots, x_n$ given the value of θ is converted into a function of θ given the observed values of $x_1, x_2, \ldots, x_n$ by (implicitly or explicitly) multiplying it by a function embodying the prior information on θ. The posterior density function obtained in this way is employed when one wishes to conduct a *Bayesian* analysis of the model.

For a non-Bayesian analysis of the same model, the function L itself is treated as if it were a function of the parameter θ given the observations $x_1, x_2, \ldots, x_n$ (rather than as a function of the observations given the parameter). In this context, the unknown parameters are estimated by choosing values for them in such a way as to maximise the likelihood function.

For example, in the special case of the normal distribution with mean θ and variance unity, we find that X has density function

$$f(x_i; \theta) = \frac{1}{\sqrt{2\pi}} e^{\frac{-1}{2}(x_i - \theta)^2}$$

and that θ has a likelihood function

$$\prod f(x_i; \theta) = \frac{1}{\sqrt{2\pi}} e^{\frac{-1}{2} \sum (x_i - \theta)^2}$$

This likelihood function is clearly maximised when the scaled sum of squared deviations

$$\frac{1}{2} \sum (x_i - \theta)^2$$

is minimised.

William Fishburn Donkin (1844) gave a mechanical interpretation of this procedure by implicitly suggesting that we may regard the negative of the logarithm of the likelihood function as if it were a potential energy function. Moreover, this interpretation of the sum of the squared deviations function as a potential energy function may be extended to a wide range of traditional optimality criteria which need not be related to probability density functions. In particular, the sum of the pth powers of the absolute deviations criterion

$$\sum |x_i - \theta|^p$$

may be interpreted as a potential energy function for all positive values of the parameter p. Further, this optimality criterion may be derived from the density function of a symmetric stable law if (and only if) the value of p lies in the range $0 < p \leq 2$.

An immediate consequence of this interpretation of the chosen optimality criterion as a potential energy function is that we may interpret the derivative of this function with respect to the parameter values as the negative of a force function acting on the current parameter values. This aspect of the subject will be developed further in later chapters.

2.1.2 Hypothesis Tests Based on Differences in Energy Levels

In Subsection 2.1.1 we have seen that the sum of squared deviations optimality function may be interpreted as a potential energy function. In general, the optimal values of the parameters are determined by the unconstrained minimum of this function. However, if the values of the parameters are known (or supposed) to satisfy some constraint (which may or may not be true) then the values of the parameters which minimise the potential energy function subject to the constraint will be associated with a value of this function which is necessarily no smaller than the unconstrained value. We may therefore use this difference in potential energy levels as the basis of a natural test of the hypothesis that the values of the parameters in the underlying statistical model truly satisfy the given constraint. For example, in the context of the linear regression models of Chapters 4 and 5, scaled variants of this difference in potential energy levels define the Wald and Lagrange multiplier statistics whereas the difference between the logarithms of the potential energy levels defines the logarithm of the conventional likelihood ratio statistic.

Figure 2.1 shows a bowl-shaped potential energy function defined as a function of the parameters b and a where b increases as we move to the right of the figure, and a increases as we move

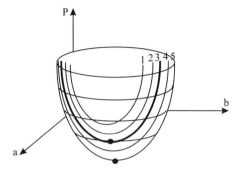

Figure 2.1   General potential energy function.

downwards and to the left. Also shown are some typical curves in the surface of this function corresponding to successively larger values of a. The overall (unconstrained) minimum of this function corresponds to the lowermost point on the fifth of these curves. If, however, the value of a is known (or supposed) to take the value represented by the third curve, then the constrained minimum corresponds to the lowermost point on this curve, and this optimal point is necessarily associated with a larger value of the potential energy function. If the Wald, Lagrange multiplier, or likelihood ratio statistic defined by the difference between these two potential energy levels is larger than an agreed critical value, then we would be inclined to reject the hypothesis that a truly takes the value implied by the third curve.

2.2   Geometrical Representations in Prediction Space

Putting aside these abstract mechanical models for the present, we turn our attention to geometrical representations of the data. There are at least three distinct geometrical representations of a set of observations; we shall refer to these three representations as lying in the prediction, observation, and parameter spaces respectively. The first two representations are more familiar to statisticians than the third; however, apart from a brief mention in

Chapter 5, we shall have no role for the prediction space representation in this book. On the other hand, we shall make considerable use of the little known parameter space representation in Chapters 5, 6, and 7.

2.2.1 Prediction Space Representation

Suppose that we have a set of n observations $y_1, y_2, \ldots, y_n$ on a single variable Y, then what we shall name the prediction space representation of this set of observations makes use of an n-dimensional Cartesian space with n mutually orthogonal axes. The first of these axes is associated with the value taken by the first observation, the second axis with the value taken by the second observation, and so on. Thus, in this representation, we identify the set of n observations on the variable Y with the single point in this space with coordinates $(y_1, y_2, \ldots, y_n)$ relative to the given system of orthogonal axes.

Now, suppose that we also have a set of n observations $x_1, x_2, \ldots, x_n$ on a second variable X, then the prediction space representation of this second set of observations makes use of the same n-dimensional space and represents this set of observations as a point in this space with coordinates $(x_1, x_2, \ldots, x_n)$ relative to the same system of axes. If each of these n observations on X is multiplied by the same factor b where $0 \leq b \leq 1$, then the result may be identified as a point on the line segment joining the origin $(0, 0, \ldots, 0)$ of the space to the point with coordinates $(x_1, x_2, \ldots, x_n)$ and lying a proportion b of the distance from the origin of this point. If the parameter b is permitted to range over all values in the interval $0 \leq b \leq 1$ then, as indicated in Figure 2.2, this set of points will trace the entire line segment between these two points. This line segment is often used as a geometrical representation of the abstract vector determined by this set of observations. (Note that we have marked the axes of Figure 2.2 with the symbols X_1, X_2, and X_3 when we should perhaps have used the symbols X_1, Y_1 on the first axis, X_2, Y_2 on the second, and X_3, Y_3 on the third, see Figure 2.3. However, in general, we prefer to avoid this multiplicity of symbols by adopting one of the relevant variables as an indicator. It would

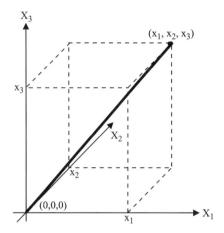

Figure 2.2   Prediction space representation of a set of observations.

perhaps have been more correct to adopt an imaginary variable as indicator and to mark the axes E_1, E_2, and E_3.)

2.2.2   A Geometrical Analogy for the Correlation Coefficient

Let O, X, and Y denote the points in n-dimensional prediction space with coordinates $(0, 0, \ldots, 0)$, $(x_1, x_2, \ldots, x_n)$ and $(y_1, y_2, \ldots, y_n)$. Further, let θ represent the angle between the line segments OX and OY, then it is well known that the squared distance between the points X and Y is related to the squared distance between the points O and X and the squared distance between the points O and Y by the following generalisation of Pythagoras's theorem:

$$\sum (x_i - y_i)^2 = \sum x_i^2 + \sum y_i^2 - 2\sqrt{\left[\sum x_i^2 \sum y_i^2\right]} \cos(\theta)$$

Now, the expression on the left of this equation may be written in the form

$$\sum (x_i - y_i)^2 = \sum x_i^2 + \sum y_i^2 - 2\sum x_i y_i$$

so that the cosine of the angle between the lines OX and OY

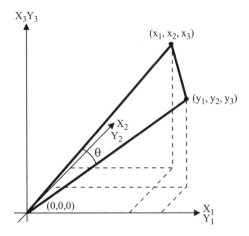

Figure 2.3 Prediction space representation of the correlation coefficient.

$$\cos(\theta) = \frac{\sum x_i y_i}{\sqrt{[\sum x_i^2 \sum y_i^2]}}$$

is a geometrical model for the correlation between the variables X and Y. The familiar formula for simple correlation

$$\cos(\theta) = \frac{\sum (x_i - \bar{x})(y_i - \bar{y})}{\sqrt{[\sum (x_i - \bar{x})^2 \sum (y_i - \bar{y})^2]}}$$

is immediately obtained if the observations on X and Y are assumed to be expressed in deviations from their arithmetic means. See Leung and Lam (1975) for a detailed account of this geometrical interpretation of the correlation coefficient, and Rousseeuw and Molenberghs (1994) for more recent developments in this area.

2.2.3 Planes and Hyperplanes in Prediction Space

In this book we shall not be concerned with models suggested by the finite line segment between the points O and X, but with those suggested by the straight line of infinite length through these points which is traced out when the parameter b of Subsection 2.2.1 is permitted to range over all values of b, that is when $-\infty < b < +\infty$.

Now suppose that we also have a set of observations on a third variable Z, then this observation may be represented by a point in the same n-dimensional space as the observations on X, and all common multiples of the observations on Z may be represented as a second line through the origin of the same space. In a similar way, the set of points determined by adding a multiple of the observations on X to a (distinct) multiple of the corresponding observations on Z determines a two-dimensional plane passing through the origin and the points determined by the sets of observations on X and Z respectively.

In a similar way, if we have sets of n observations on q explanatory variables $X_1, X_2, \ldots, X_q$ then these observations determine a set of q points in n-dimensional space and, in general, the set of all linear combinations of these q sets of observations determines a q-dimensional plane or hyperplane passing through the q given points and the origin. (In passing we note that some authors prefer to reserve the word *hyperplane* for the special case in which the

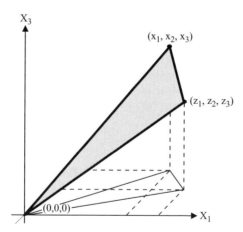

Figure 2.4   Prediction space representation of linear combinations of two sets of observations.

dimension of the subspace is one less than the dimension of the whole space, that is when $q = n - 1$.)

We shall defer our discussion of the prediction space representation of the data until after we have examined the relevant line fitting procedures in Chapter 5. Even then, our account of the subject will be brief and we refer readers to Durbin and Kendall (1951) for further details. In this context, it should be noted that they refer to our "prediction space" as "observation space."

2.3 Observation Space Representation

In the previous section, we have seen that the prediction space representation of the n observations on a set of q explanatory variables and a single dependent variable is as a set of $q + 1$ points in n-dimensional space. By contrast, what we shall call the observation space representation of these n observations is as a set of n points in $(q + 1)$-dimensional space. If this set of observations exactly satisfies the relationship

$$y_i = x_{i1}b_1 + x_{i2}b_2 + \ldots + x_{iq}b_q \quad i = 1, 2, \ldots, n$$

for some sets of values $b_1, b_2, \ldots, b_q$, then these n points will lie on a q-dimensional hyperplane in this $(q + 1)$-dimensional space.

Further, if all the observations on one of the variables, say the first, take the same value, say unity, then the dimensions of both the space itself and the specified hyperplane may be reduced by unity. Thus, if this set of observations exactly satisfies the relationship

$$y_i = b_1 + x_{i2}b_2 + \ldots + x_{iq}b_q \quad i = 1, 2, \ldots, n$$

for some set of values $b_1, b_2, \ldots, b_q$, then they will lie on a $(q - 1)$-dimensional hyperplane in q-dimensional space.

Restricting our discussion to the case $q = 2$ with $x_{i1} = 1$, we find that the ith observation $X_1 = 1, X_2 = x_i$, and $Y = y_i$ on the variables X_1, X_2, and Y may be represented by the point (x_i, y_i) in the two-dimensional Cartesian plane with its x-axis running from west to east along the line $y = 0$ and its y-axis running from south to north along the line $x = 0$. Further, we note that all observations satisfying the relationship

$$y_i = a + bx_i \quad i = 1, 2, \ldots, n$$

will lie on a (one-dimensional) straight line in this two-dimensional plane. This situation is illustrated in Figure 2.5.

The point $(x, y) = (4, 6)$ is also plotted in this figure, for reasons to be explained below.

2.4 Parameter Space Representation

Again restricting our discussion to the case $q = 2$ with $x_{i1} = 1$, we find that we may also represent the set of all observations satisfying the relationship

$$y_i = a + bx_i \quad i = 1, 2, \ldots, n$$

by the values of the parameters a and b or, more formally, we may represent this set of observations by the point (b, a) in the two-dimensional Cartesian parameter plane with its b-axis running from west to east along the line $a = 0$ and its a-axis running from south to north along the line $b = 0$. Thus a line in the xy-

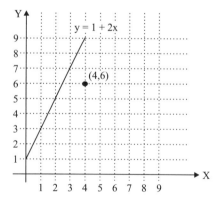

Figure 2.5 Observation space representation of a single observation and a linear relationship between a set of observations.

plane corresponds to a point in the ba-plane. (Note that, for our later convenience, we have reversed the lexicographical ordering of the letters a and b here. However, there is no formal reason why we cannot employ these letters in their usual ordering should we wish to do so.)

This situation is illustrated in Figures 2.5 and 2.6 as the line $y = 1 + 2x$ (with parameters $a = 1$ and $b = 2$) in the xy-plane of Figure 2.5 may be represented by the point $(b, a) = (2, 1)$ in the ba-plane of Figure 2.6

Now, consider what is implied by the line

$$a = y_0 - x_0 b$$

in the ba-plane. It is clear that this equation defines the set of intercepts a and slopes b of the family of lines that pass through the point (x_0, y_0) in the xy-plane. This situation is illustrated in Figure 2.7.

We note that the points $(0.0,6)$, $(0.5,4)$, $(1.0,2)$, and $(1.5,0)$ all lie on the line $a = 6 - 4b$ in the ba-plane of Figure 2.7. The lines in the xy-plane $y = 6$, $y = 4 + 0.5x$, $y = 2 + x$, and $y = 1.5x$ corresponding to these four points in the ba-plane all pass through the point $(x, y) = (4, 6)$ in the xy-plane of Figure 2.8, and the relationship between the line $a = 6 - 4b$ in the ba-plane and the point $(x, y) = (4, 6)$ in the xy-plane is established.

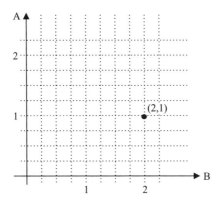

Figure 2.6 Parameter space representation of a linear relationship between observations.

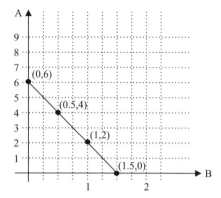

Figure 2.7 Parameter space representation of a linear relationship between parameters.

Thus, in the same way as a line in the xy-plane corresponds to a point in the ba-plane, a line in the ba-plane will correspond to a point in the xy-plane.

However, there is a subtle difference, as may be seen by combining Figures 2.6 and 2.7 to form Figure 2.9 before comparing the line $a = 6 - 4b$ and the point $(b, a) = (2, 1)$ in the ba-plane of Figure 2.9 with the point $(x, y) = (4, 6)$ and the line $y = 1 + 2x$ in the xy-plane of Figure 2.5. Clearly, a point in the positive quad-

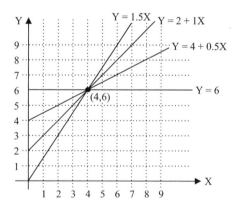

Figure 2.8 Observation space representation of a linear relationship between parameters.

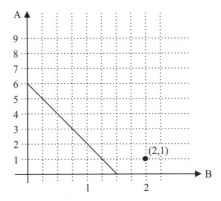

Figure 2.9   Parameter space representation of a set of parameters and a linear relationship between sets of parameters.

rant of the xy-plane defines a line in the ba-plane with a negative slope, whilst a point in the positive quadrant of the ba-plane defines a line in the xy-plane with a positive slope. Indeed, the particular form of this relationship establishes that the parameter space representation of a fitting problem is the projective geometry dual of the observation space representation of the same problem and *vice versa*.

2.5   Line Fitting by Eye

As an exercise in the use of these last two representations, and as a preparation for our later discussion of the line fitting problem, the reader is invited to consider the problem of choosing a line which seems best to fit the $n = 5$ observations on X and Y given in Table 2.1. In the present context, this problem is to be solved without the aid of an explicit optimality criterion.

As explained in Section 2.3, the artificial data given in Table 2.1 may be plotted as a scatter of points in the xy-plane, see Figure 2.10. Using the edge of a transparent ruler or a taut length of black cotton, we may adjust the position of the straight line defined in this way until it determines a line which seems to fit this set of observations as closely as possible (where, for the time being, it is

Table 2.1 Simple numerical example

X	Y
2	2
3	6
4	4
6	8
8	4

left to the reader to decide what is meant by the phrase "as closely as possible").

The results obtained here without the benefit of an explicit optimality criterion may be compared with the lines fitted by two variants of the method of least squares illustrated in Figures 1.1 and 1.2.

This traditional line fitting technique may readily be automated by using a computer mouse to adjust the positions of the midpoint and one end of the candidate fitted line in the xy-plane. This is the technique recommended by Bajgier, Atkinson, and Prybutok (1989). However, their proposal involves the use of a

Figure 2.10   Observation space representation of the data in Table 2.1.

single computer mouse for two distinct purposes. It is more convenient to employ the mouse to identify suitable values for the slope and intercept parameters of the candidate line directly in the *ba*-plane. In principle, we have to display two planes on the same monitor screen, one to represent the positions of the observed points and the fitted line in observation space, and the other to represent the values of the slope and intercept parameters of the fitted line in parameter space. The desired result may either be achieved by splitting the monitor screen into two portions, or by employing a toggle to alternate between the two representations of the problem.

We shall not develop a computer program of the type outlined here. But, supposing that we had done so, then the practitioner would have been asked to use the computer mouse in the *ba*-plane to determine the position of the point corresponding to the slope and intercept parameters of the candidate fitted line in the *xy*-plane. The position of the line which seems best to fit the given observations would then be determined by visual inspection.

At the conclusion of the fitting process we might have a pair of screens such as those indicated in Figure 2.11.

This empirical determination of a supposedly optimal fitted line suffers from the obvious defect that the practitioner's choice is made without reference to an explicit goodness of fit criterion. If it is thought appropriate to employ the least squares criterion of Chapter 4 in this role, then this criterion may readily be incorpo-

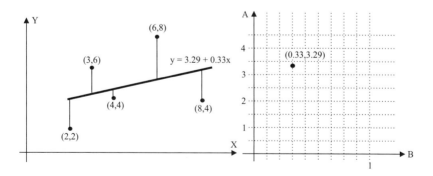

Figure 2.11   Dual representations of the line fitting problem.

rated into the proposed package by supposing that the procedure also returns values for the functions $\sum_i e_i$ and $\sum_i x_i e_i$ where $e_i = y_i - a - bx_i$ is the deviation between the ith observation and the fitted line when distances are measured parallel to the y-axis. In this context, the practitioner would be asked to adjust the position of the computer mouse in the ba-plane in such a way as to bring the values of these functions as close to zero as possible. The resulting estimates of the slope and intercept parameters are not inherently better than those obtained by visual inspection, but they do correspond to the values generated by a widely used fitting procedure.

References

Bajgier, S. M., M. Atkinson, and P. B. Prybutok (1989), Visual fits in the teaching of regression concepts, *The American Statistician* **43**: 229–233.

Donkin, W. F. (1844), An essay on the theory of combination of observations, *Transactions of the Ashmolean Society* **2** (18):1–71.

Durbin, J. and M. G. Kendall (1951), The geometry of estimation, *Biometrika* **38**: 150–158.

Farebrother, R. W. (1992), Geometrical foundations of a class of estimators which minimise sums of Euclidean distances and related quantities, in Y. Dodge (Ed.), *L₁-Statistical Analysis and Related Methods*, North-Holland Publishing Company, Amsterdam, 337–349.

Leung, C. K. and K. Lam (1975), A note on the geometric representation of the correlation coefficients, *The American Statistician* **29**: 129–130.

Rousseeuw, P. J. and G. Molenberghs (1994), The shape of correlation matrices, *The American Statistician* **48**: 276–279.

CHAPTER 3

Mechanical Models for Multidimensional Medians

3.1 Mechanical Models for One-Dimensional Medians

3.1.1 Potential Energy Models for One-Dimensional Medians

As a concrete application of the abstract considerations discussed in Chapter 2, we anticipate the work of Chapter 5 to a certain extent by developing mechanical models for medians of sets of data in one and two dimensions based on systems of strings and pulleys.

In Chapter 1, we discussed the possible use of a chemical balance in a state of equilibrium as a model for the arithmetic mean of a set of n direct observations on a single variable. Unfortunately, this model is not suitable for our purposes as it cannot readily be extended to the median and other one-dimensional means, as it does not specify a criterion function that is to be minimised, but only a balance of rotational forces which must hold when the system is in equilibrium.

Our first problem is thus to specify the potential energy function associated with the value taken by a single observation. Suppose that we are given a horizontal board. Then we may draw a straight line on it and identify this line with the y-axis of a Cartesian system. Arbitrarily selecting a point on this line, we may drill a hole through the board at this point and pass a length of string through the hole. We attach a unit weight to the lower end of the string which thus hangs vertically from the hole whilst the portion of the string which is above the board is pulled taut along the given line. As the upper end of the string is pulled away from the hole along this line the weight will rise and thus gain potential energy at a steady rate. On the other hand, if the upper end of the string is allowed to move towards the hole, then the weight will fall and lose potential energy at the same steady rate. More precisely, if the hole is at the point $y = y_i$ on the y-axis and the upper end of the string is at the point $y = a$, then the distance between these two points is $|y_i - a|$. Further, since the upper end of the string may be regarded as having moved this distance from the hole, the unit weight will be at the same distance above its lowest point, and hence the potential energy level associated with this simple model will be proportional to the function

$$P_i = |y_i - a|$$

Now, our discussion of this model is in the context of a given data set, so that the ith observation $Y = y_i$ may be regarded as fixed and the potential energy function P_i regarded as a function of the variable parameter a.

Thus, we have to plot this function in the vertical plane with a (rather than y) on the horizontal axis and P_i on the vertical axis. In this context, and employing a natural scaling, we find that the potential energy function P_i takes the form of two half-lines rising

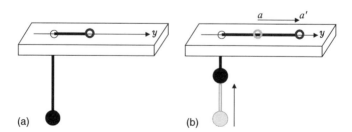

Figure 3.1 Simple mechanical model (a) initial position, and (b) subsequent position.

in the vertical plane at angles of 45 degrees from the point $a = Y_i$ on the horizontal a-axis, see Figure 3.2. In later diagrams we shall find it convenient to rescale the P-axis to yield functions with gentler slopes.

Now suppose that, instead of a single point on the y-axis, we have a scatter of n points corresponding to a set of n observations $y_i, y_2, \ldots, y_n$ on a single variable Y. Again identifying the y-axis with an arbitrary line in the horizontal plane, we may drill holes through the horizontal board at each of the n locations corresponding to observations and pass a length of string through each of these holes. We again attach unit weights to the lower ends of these strings and tie their upper ends to a single ring at an arbitrary point $y = a$ on the y-axis.

For clarity in Figure 3.3a, we have elongated the circular ring into an oval with straight sides and separated the y-observations along the major axis of this oval. More formally, we have associated a set of arbitrary x-values with the given y-values and replace the ring by a rigid rod that is constrained to move along a pair of tracks that run parallel to the y-axis. As the ring is moved along the y-axis, the weights rise and fall as before, thus defining the potential energy functions of the n individual weights, see Figure 3.3b. The sum of these individual functions defines the piecewise linear potential energy function of the system as a whole,

$$P = \sum_i |y_i - a|$$

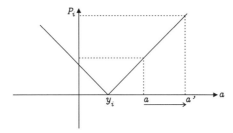

Figure 3.2   Potential energy function of a single observation.

Now, if the observations on Y are strictly ordered, $y_1 < y_2 < \ldots < y_n$, then, for values of a less than y_1, the overall potential energy function is given by

$$P = y_n + \ldots + y_3 + y_2 + y_1 - na$$

with a slope of $-n$. Similarly, for values of a between $a = y_1$ and $a = y_2$, this function is given by

$$P = y_n + \ldots + y_3 + y_2 - y_1 - (n-2)a$$

with a slope of $2 - n$. Again, for values of a between $a = y_2$ and $a = y_3$, this function is given by

$$P = y_n + \ldots + y_3 - y_2 - y_1 - (n-4)a$$

with a slope of $4 - n$, and so on. Thus, the $n + 1$ line segments that constitute the piecewise linear overall potential energy function have gradually increasing slopes of $-n, 2 - n, \ldots, n - 2, n$ that are joined to one another at the points $a = y_1, a = y_2, \ldots, a = y_n$, again see Figure 3.3b.

Further, the value of a which minimises the overall potential energy function defines a measure of central tendency which, in Subsection 3.1.2, we shall identify as the median or middlemost

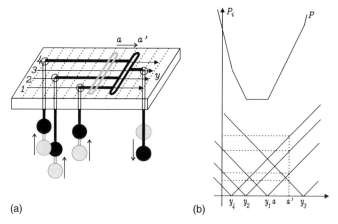

(a)   (b)

Figure 3.3   (a) Balance of forces, and (b) the potential energy functions associated with a set of observations

value of the set of observations. In general, we would attempt to determine this optimal value of a by finding the value of a at which the derivative of the function is equal to zero. However, this standard approach to the problem cannot be employed in the present context as the derivative of this piecewise linear function takes a maximum of $n + 1$ distinct values corresponding to the n $+1$ distinct slopes of its constituent line segments. Moreover, it is possible that none of the $n + 1$ values taken by the derivative will be zero. It is therefore necessary, for the present, to suppress the potential energy function and to adopt an alternative approach based on the balance of forces in the strings of the model. We shall, however, return to a more explicit potential energy analysis of this and related problems in later chapters.

3.1.2 Balance of Forces in One-Dimensional Medians

Consider the mechanical model of Subsection 3.1.1. Let s_i represent the force in the ith string tending to pull the ring in the direction of increasing values of y, then this function will take the value $s_i = +1$ if y_i is greater than a and the value $s_i = -1$ if y_i is less than a. For the present, we shall ignore the case when y_i is equal to a.

In this context, we may identify the derivative of the ith potential energy function with the negative of this force function,

$$\frac{dP_i}{da} = -s_i$$

at least when a is not equal to y_i. We may therefore examine the minimisation problem on the basis of the balance of forces in the strings of the system. For ease of exposition, we shall suppose that the y-axis runs from south to north through the horizontal plane. If there are r observations to the south of the current value and $n - r$ to the north of this value, then there will be $n - r$ strings pulling the ring northwards and r strings pulling it southwards. Thus, there is a net force of $n - 2r$ units pulling the ring northwards and the ring will move in this direction if n is greater than $2r$ and in the opposite direction if n is less than $2r$. If n is an odd number satisfying $n = 2m + 1$ then the ring will be at rest if it is located at the $(m + 1)$th observation. Similar, if n is even and $n = 2m$ then the ring will be at rest at any point on the interval between the mth and the $(m + 1)$th observations. It is well known that these two properties characterise the *median* of a set of uni-variate observations. We may therefore deduce that the measure of central tendency defined by the value of a which minimises the piecewise linear potential energy function P of Subsection 3.1.1 is the median of the given set of n observations.

In passing, we note that these results give some indication of possible values of the force function s_i when $a = y_i$, for s_{m+1} takes the value 0 in the first case when $a = y_{m+1}$ and s_m takes the value -1 and s_{m+1} takes the value $+1$ in the second case even though a may take the values y_m or y_{m+1} as well as any value between the two. In fact, it can be shown that s_i may take any value in the range $-1 \leq s_i \leq +1$ when $a = y_i$.

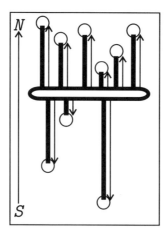

Figure 3.4   Balance of forces in the south-north direction.

3.2 Mechanical Models for Two-Dimensional Medians

3.2.1 Potential Energy Models for Two-Dimensional Medians

Now suppose that, instead of a scatter of points on the horizontal y-axis, we have a scatter of points in the horizontal xy-plane. We select an arbitrary point in this plane, and seek to choose this point in such a way that the sum of the absolute distances from the n given points to this additional point is a minimum. The measure of central tendency determined in this way is known as the *mediancentre* or the centre of population as opposed to the conventional *centroid* or centre of gravity which minimises the sum of the squared distances instead of the sum of their absolute values.

To construct a mechanical model for the mediancentre, we again drill holes through a horizontal board at the Cartesian points (x_i, y_i) corresponding to the n observations on X and Y. We attach weighted strings running from the given points to a ring at the arbitrary point (c, a). These unit weights induce unit forces tending to pull the ring from its present position towards the n holes representing the n individual observations. Resolving the n forces acting on the ring parallel to the x-axis and parallel to the y-axis and summing the results, we obtain two expressions which must take zero values if the ring is to be in a state of equilibrium. The values of $x = c$ and $y = a$ associated with this state of equilibrium determine the position of the mediancentre.

This mathematical problem was first defined by Pierre Fermat in 1648. The corresponding mechanical model was subsequently developed by Pierre Varignon in 1687 and by Gabriel Lamé and Benoit Paul Emile Clapeyron in 1829. This mechanical model is also associated with the name of Alfred Weber although the section of his book of 1909 in which this model was developed was actually written by his colleague Georg Pick; see Franksen and Grattan-Guinness (1989) for a more detailed discussion of the history of this problem.

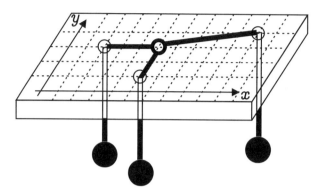

Figure 3.5   Mechanical model in the two parameter case.

Lamé and Clapeyron were more concerned with the elimination of friction than we are; they wished to use their model as an
analogue computer for obtaining precise estimates of the correct
solutions of the fitting problem, whereas we are only concerned
with developing a satisfactory mechanical representation of a statistical problem that we can solve by numerical means. Thus, in
their model, Lamé and Clapeyron did not drill holes through a
horizontal board but attached pulleys mounted on spindles to the
corresponding positions on the underside of this board and passed
strings over these pulleys to obtain a mechanical model which is
less liable to friction and thus better able to produce an accurate
approximation to the required solution, see Figure 3.6.

3.2.2   Geometrical Solution in Parameter Space

In principle, we may solve the problem of fitting a mediancentre to
a set of direct observations on two variables by a geometrical
procedure. In the context of a one-dimensional plot of direct observations we were able to construct an explicit objective function by
plotting the V-shaped error functions in respect of each observation in a second dimension at right angles to the first. These individual error functions are then summed and the optimal value of
the parameter a determined by the minimum value of this function.

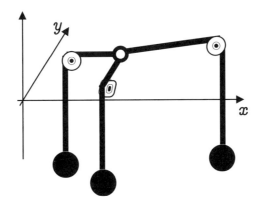

Figure 3.6 Two-dimensional mechanical model based on strings and pulleys.

In the context of the two-dimensional problem, the V-shaped functions of Subsection 3.1.1 have to be replaced by circular cones with V-shaped cross-sections in all directions. Thus we have to construct a cone of this type at each of the n points in the horizontal ca-plane and add the resulting functions together to form the overall optimality criterion. If necessary, this system of cones may be represented by a system of circular contours. However, as neither of these schemes is particularly helpful to our understanding of the problem, we shall not trouble to illustrate them here.

In passing, we note that the same system of circular contours may be used to represent the more general distance functions discussed in Chapters 4 and 5. Indeed, this approach may be applied when the distance between an observed point and an arbitrary point is any increasing function of the Euclidean distance between these two points. The only difference being that the successive contours are associated with different values when alternative measures of distance are used.

3.2.3 Vectorial Representation of Forces

Turning now to a more algebraic treatment of the problem, we note that force vectors may be resolved parallel to each member of a given set of orthogonal axes and then reconstructed from their constituent parts. This result may be familiar to some readers under the name of the "parallelogram of forces rule".

Suppose that we are given a vector of length w pointing in a direction which is θ degrees to the east of north. Then the angle between north and the vector is θ degrees and we may replace the single force acting in this direction by a vector of length $w\cos(\theta)$ acting in a northwards direction and a vector of length $w\sin(\theta)$ acting in an eastwards direction. Conversely, if we have a vector of length $f = w\cos(\theta)$ acting in a northwards direction and a vector of length $g = w\sin(\theta)$ acting in an eastwards direction then we may replace these two vectors by a single vector of length $w = \sqrt{(f^2 + g^2)}$ acting in a direction which is at an angle $\theta = \arctan(g/f)$ to the east of north.

To establish the effect of a set of n forces acting on a ring, we consider each of the forces in turn, resolve them into their constituent parts acting in a northwards direction and in an eastwards direction. The n parts acting in a northwards direction are then summed to determine the overall force in this direction. The resolved forces acting in an eastwards direction are similarly summed to obtain the overall force acting in that direction. Finally, these two constituent parts are recombined to determine the direction and magnitude of the net force acting on the ring.

This algebraic analysis may readily be given a geometrical interpretation. A force of magnitude w_i acting in a direction θ_i

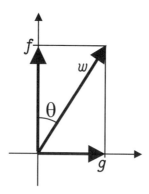

Figure 3.7 A force vector resolved into its orthogonal components.

degrees to the east of north is conventionally represented as an arrow of length w_i pointing in this direction. Given a set of n such arrows representing the n forces acting on the ring, we may perform the calculations outlined in the previous paragraph by taking each of the forces in turn, placing the tail of each vector at the head of the previous vector so that the northwards and eastwards components are summed automatically. Finally, we have to join the tail of the first vector to the head of the last vector to determine the magnitude and direction of the net force acting on the ring. In this context, it should be noted that the same result is obtained independently of the order in which the vectors are processed.

For example, if we are given a set of three observations $(x_1, y_1) = (0, 5)$, $(x_2, y_2) = (5, 0)$ and $(x_3, y_3) = (8, 9)$, then these observations may be represented by three points at the locations (0,5), (5,0) and (8,9) in the Cartesian plane. Taking $(c, a) = (5, 5)$ for our trial value, we find that these observations may also be represented by points at locations $(-5, 0)$, $(0, -5)$, and $(3, 4)$ when the origin of the Cartesian system has been removed to the trial value. That is, the first observation is 5 units to the west of the trial value, the second is 5 units to the south, and the third is 3 units to the east and 4 to the north. Rescaling the vectors implied by these expressions to unit length, we have $(-1, 0)$, $(0, -1)$, and $(0.6, 0.8)$ for the unit forces acting on a ring at the trial position, see Figure 3.8a. That is, the first force pulls the ring westwards, the second pulls it southwards, and the third pulls it in a direction which is arctan$(0.75) = 36° 52'$ to the east of north. Summing these unit forces, we have the expression $(-0.4, -0.2)$ for the overall force acting on the ring. Thus this force has magnitude $\sqrt{0.20} = 0.45$ and it acts in a direction which is arctan$(2.0) = 63° 26'$ to the west of south.

Two distinct geometrical constructions leading to the same numerical results are illustrated in Figures 3.8b and 3.8c.

As a second instance of this geometrical technique, suppose that we have a set of three unit vectors that may be rearranged to form a closed equilateral triangle. For example, suppose that we have a vector running from the point (0, 0) to the point $(0, -1)$, a second vector running from the point $(0, -1)$ to the point $(\sqrt{3}/2, -1/2)$, and a third vector running from the point $(\sqrt{3}/2,$

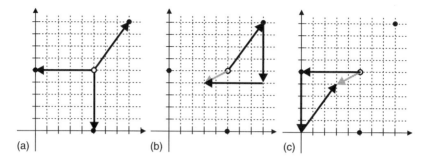

Figure 3.8 (a) Three forces acting on a point; (b) and (c) two determinations of the associated net force.

−1/2) to the point (0,0). Since these three vectors form a closed equilateral triangle, there must be a net force of zero acting on the ring.

To demonstrate this point in an alternative manner, we suppose that one of the unit vectors points due south and that the other two point 60 degrees east of north and 60 degrees west of north respectively, see Figure 3.9a. Now, we may replace the arrow of length unity pointing north 60 degrees east by an

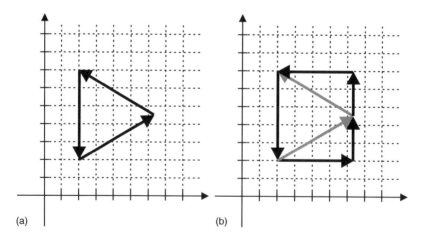

Figure 3.9 (a) Geometrical representation of three forces in equilibrium, and (b) balance of the corresponding resolved forces.

arrow of length $\sin(60°) = \sqrt{3}/2$ pointing due east and an arrow of length $\cos(60°) = 1/2$, pointing due north. Similarly, the arrow of unit length pointing north 60 degrees west may be replaced by an arrow of length $\sin(60°) = \sqrt{3}/2$ pointing due west and an arrow of length $\cos(60°) = 1/2$ pointing due north. In each case the original unit vectors may be reconstructed from these solutions. Now, the vector pointing due east exactly balances that pointing due west, and the sum of the two vectors pointing north exactly balance the unit vector pointing south, so that the net force tending to move the ring from its present position is zero, see Figure 3.9b.

Writing the three vectors in this second example in their resolved forms and multiplying by $\sqrt{2/3} = 2/\sqrt{6}$ we have the scaled system of vectors:

$$(1/\sqrt{2}, 1/\sqrt{6}), (-1/\sqrt{2}, 1/\sqrt{6}), (0, -2/\sqrt{6})$$

where the first element of each pair indicates the distance in an eastwards direction and the second the distance in a northwards direction.

Arranging these vectors as a matrix and removing square roots, it will be clear to readers familiar with the Helmert transformation that the corresponding system of vectors for the three-dimensional problem are given by

$$(1/\sqrt{2}, 1/\sqrt{6}, 1/\sqrt{12})$$
$$(-1/\sqrt{2}, 1/\sqrt{6}, 1/\sqrt{12})$$
$$(0, -2/\sqrt{6}, 1/\sqrt{12})$$
$$(0, 0, -3/\sqrt{12})$$

each of which has to be multiplied by a factor of $\sqrt{4/3}$ to restore them to unit length.

Finally, we note that the vectors in this three-dimensional system may be associated with the vertices of a regular tetrahedron or triangular pyramid.

3.2.4 Particular Solutions

As in the one-dimensional case, we are again able to obtain explicit general solutions for the unrestricted problem in certain special cases. In particular, if there are just three observations and the largest angle of the triangle defined by these observations is less than 120 degrees, then a consideration of the relevant forces reveals that the solution will lie at the interior point of the triangle which subtends an angle of 120 degrees with each pair of observed points, see Figure 3.10a. On the other hand, if the largest angle of the triangle defined by the observations is 120 degrees or more, then the solution will be at this largest angle as the resolved force pulling the ring away from this point is necessarily no larger than the unit force tending to hold it in position, see Figure 3.10b.

The solution in this special case is thus characterised by a set of three unit lines pointing towards the corners of a plane equilateral triangle. This characterisation of the solution may readily be generalised to the case when there are $q + 1$ observations in q-dimensional space. For example, if there are four observations in

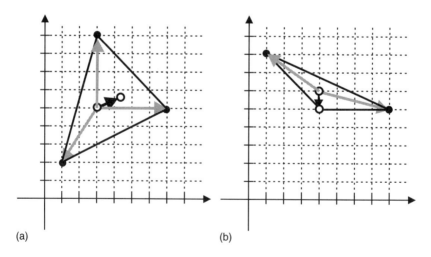

(a)   (b)

Figure 3.10 Particular solution when there are three observations in two dimensions, (a) when all angles are less than 120°, and (b) when the largest angle is greater than 120°.

three-dimensional space, then the solution is characterised by a set of unit lines pointing towards the four corners of a regular triangular pyramid.

Returning to the two-dimensional problem, we note that if there are four observations in the form of a convex quadrilateral, then the solution will lie at the point in the interior of this quadrilateral corresponding to the intersection of its diagonals. In this position the unit forces tending to pull the ring towards each of a pair of diagonally opposite points will exactly balance that tending to pull it in the opposite direction, see Figure 3.11. Note that the quadrilateral defined by the four points must be convex for this characterisation of the solution to be valid.

3.2.5 Algorithm for the Mediancentre

Our analysis of the mechanical model for the mediancentre of a set of observations relates to the balance of forces when the system is in a state of equilibrium. By definition, it is necessarily a static analysis. To obtain a computational algorithm based on this model we must explain how the ring moves when the system is

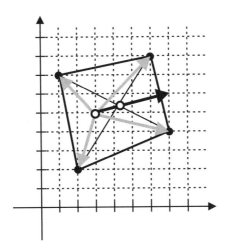

Figure 3.11 Particular solution when there are four observations in two dimensions.

not in a state of equilibrium. The following analysis is not strictly dynamic but what physicists call "quasi-static" and what economists call "comparative statics". We begin by choosing an arbitrary point in the horizontal plane. We then determine the net force acting on a ring at this point, and move a small distance in the direction indicated by the net force and proportional to its magnitude. We then recompute the net force acting on the ring in this new position, and so on until a satisfactory level of convergence is obtained. This simple vectorial analysis forms the basis of Gower's (1974) algorithm for determining the mediancentre of a set of n points in q-dimensional space with $q \geq 2$.

In Figure 3.12, suppose that the initial position of the ring is towards the right of the diagram. This ring gradually moves downwards and towards the left until it arrives at its optimal position.

In the particular instance of the first numerical example of Subsection 3.2.3, we found that the initial trial position (5, 5) is associated with a force vector of $(-0.4, -0.2)$. We have therefore to move the ring a distance proportional to $(-0.4, -0.2)$ from its initial position (5, 5). For example, we may move the ring to the point (4.6, 4.8). Relative to a Cartesian system with its origin at this

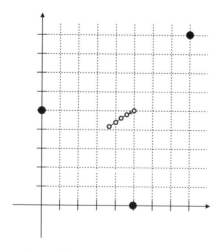

Figure 3.12 Successive positions in the iterative determination of the mediancentre.

new position, the observations may be represented by a set of three points at the locations $(-4.6, 0.2), (0.4, -4.8)$, and $(3.4, 4.2)$. The unit forces acting on a ring at this location are given by $(-0.999, 0.043), (0.083, 0.997)$, and $(0.629, 0.777)$, and the sum of these forces by $(-0.287, -0.177)$. Thus, at the second stage of this iterative procedure, we have to move a distance proportional to $(-0.287, -0.177)$ from the second trial position $(4.6, 4.8)$ to a new trial position at $(4.313, 4.623)$. If the constant of proportionality employed by this iterative procedure is sufficiently small, then it will eventually converge to a position in which there is a zero net force acting on the ring. The position identified in this way clearly defines the mediancentre of the observations.

3.2.6   Non-unit Weights

In our previous discussion we assumed that unit weights had been attached to each of the n strings in the mechanical model. Suppose that there are m distinct observations and that the ith observation is repeated a total of n_i times, then, in our basic model, there will be n_i strings each carrying a unit weight connecting the ring to the ith observation. It is immediately apparent that we may simplify this basic model by replacing these n_i strings each carrying a unit weight by a single string carrying n_i unit weights, or equivalently, by a single string carrying a single weight of n_i units.

    If the weights attached to the observations are changed, then we obtain a second weighted mediancentre which is appropriate as a measure of central tendency in the changed circumstances. For example, in their military application, Lamé and Clapeyron determined an appropriate position for the headquarters of an army corps and a distinct location for its supply depot. In the first case the unweighted sum of the distances from the headquarters to the locations of the individual staff officers is the criterion of interest, whilst in the second case the distances to the troop cantonments have to be weighted by the physical weight of the supplies that are to be delivered to these locations.

    If one of the weights in this model is sufficiently large, then the force required to remove the ring from the position corresponding to this weight cannot be achieved by combining the weights attached to the remaining observations. In this context, the optimal position of the ring is clearly determined by the location of the single observation with very large weight.

3.3 Linear, Curvilinear, and Regional Constraints

3.3.1 Linear and Curvilinear Constraints

Thus far in this chapter, we have implicitly assumed that we have to choose a value for the centre of population without any constraint on its range of possible values. A more careful formulation of the problem may suggest that the optimal point must lie on a particular straight or curved line in the plane. We then have to choose the position of a point on this line in such a way as to minimise the sum of the absolute distances from the observed points. Once again, the mechanical model takes the same general form as before except that the ring is now constrained to run along a wire in the horizontal plane, see Figure 3.13. When the ring is in equilibrium, the force pulling it in one direction along the wire will exactly balance the force pulling it in the opposite direction. Thus,

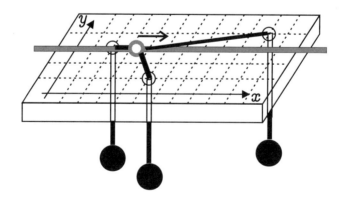

Figure 3.13 Mechanical representation of the constrained optimisation problem.

the net force acting on the ring must be resolved perpendicularly to the wire and tangentially to the wire; the first of these expressions should be ignored as not relevant to the solution of the problem, and the second set equal to zero.

If the constraining wire is curved then there is a possibility that the point of equilibrium determined in this way is a local minimum of the function. It is therefore necessary to identify all the points of equilibrium by starting the ring from a range of initial positions on the wire as illustrated in Figure 3.14.

An interesting special case of the basic problem occurs when there are just two observations and we have to locate a point on a given straight line which minimises the sum of the distances from this point to the two given points. If the two points are on opposite sides of the given line then any point on the straight line joining these two points will minimise the sum of distances optimality function.

A less trivial problem occurs when the two points are on the same side of the given line. Resolving the forces in the stretched strings perpendicularly to the line and parallel to it, we immedi-

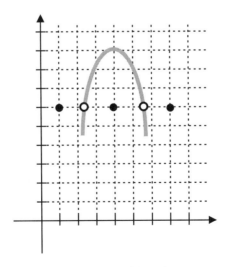

Figure 3.14   Multiple solutions identified by using a range of starting values.

ately establish that the two strings must meet the line at the same angle, see Figure 3.15. This result may be identified with the principle of reflection of light in a mirror.

Similarly, if the observed points are on opposite sides of the given line but are now associated with distinct weights, then we have to choose a point on the given line to minimise the weighted sum of the absolute distances. The situation illustrated in Figure 3.16 may be identified with the principle of refraction of light at the boundary between two translucent media. As a light beam strikes the boundary between two media, it bends in such a way that the sine of the angle of incidence divided by the sine of the angle of refraction is proportional to the ratio of the densities of the two media. In this context, the principle that the sum of the weighted distances should be minimised is known as the "principle of least work".

This discussion of reflection and refraction at a boundary may readily be generalised to the corresponding problems with curvilinear constraints. However, we shall not trouble to illustrate this more general situation.

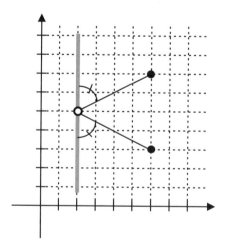

Figure 3.15 Constrained minimisation problem similar to reflection of light in a mirror.

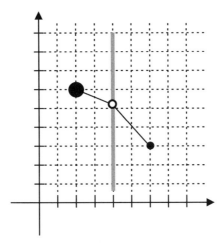

Figure 3.16 Constrained minimisation problem similar to refraction of light at a boundary.

3.3.2   Regional Constraints

Returning to the simpler model, we may constrain the ring to lie in a closed region rather than on a line. If the unconstrained optimal point lies within this region then restricting the ring to the region will impose no constraint on it, see Figure 3.17.

On the other hand, if the unconstrained minimum of the sum of distances function lies outside the given region, then the constrained optimal point will necessarily lie on the boundary of this region and we only have to search the relevant section of the boundary to determine the point which minimises the optimality function subject to this constraint. Figure 3.18 illustrates the situation in which the unconstrained minimum lies to the south and east of the shaded region so that the constrained optimal value lies on the southeastern boundary of this region.

The idea of a regional constraint is perhaps more familiar in the geometrical form of linear programming. In the two-dimensional case we are given a set of linear inequality constraints on the values of the parameters a and b. Our problem is to choose values for these parameters in such a way as to minimise the distance

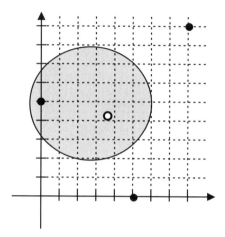

Figure 3.17 Regional constraint when the unconstrained solution lies in the shaded region.

from the chosen point to an arbitrarily remote straight line. This situation is illustrated in Figure 3.19 where we wish to maximise an objective function of the form $a + b$ subject to the constraints $a \leq 5$ and $b \leq 5$.

3.3.3   Hypothesis Tests

In earlier sections, we have seen that the sum of distances optimality function may be interpreted as a potential energy function, and that the constrained optimal point will be associated with a larger value of this function than is the unconstrained point. It therefore seems reasonable to suggest that we may use this difference in potential energy levels as the basis of a natural test of the hypothesis that the true value of the parameter in the underlying model satisfies the given constraint. Although tests based on differences in potential energy levels are indeed employed in the sum of squared differences context of Chapter 4, those used in the sum of absolute differences context of Chapters 3 and 5 are usually based on expressions that are closely related to formulas developed for the sum of squared differences case, see Subsection 5.2.4 for detailed references.

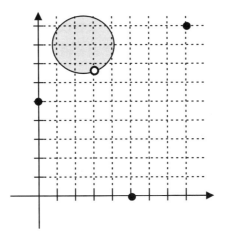

Figure 3.18 Regional constraint when the unconstrained solution lies outside the shaded region.

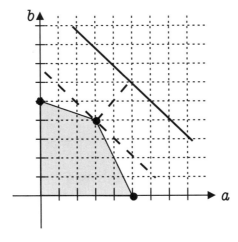

Figure 3.19 Constrained region defined by a set of linear inequality constraints.

3.4   Three Further Generalisations of the Mechanical Model

3.4.1   Models with Non-Euclidean Distances

In our discussion of the constrained problem in Section 3.3, we assumed that the point of interest must be located on a straight or curved line and that distances are measured as straight line distances between pairs of points. If, on the other hand, all distances are to be measured along specific paths in the plane, then a suitable mechanical model may be obtained by erecting tall hedges on either side of each of these paths and choosing a position for the ring in such a way that the sum of the distances measured along these paths is minimised. This simple mechanical model is again due to Lamé and Clapeyron. Our outline of the theory of metric graphs in Chapter 8 will be based on a slightly more general version of this model.

A second variant of the basic problem occurs when the solution is restricted to one of a set of specified points, and the distance from each of these specified points to each of the given points is known. The problem of choosing one of the specified points in such a way as to minimise the sum of the distances from the given points may easily be realised as a problem in integer programming, see Vinod (1969) for details.

A third variant of the basic problem is obtained if, instead of restricting the ring to lie on a given curve, we permit each of the n observations to occupy any point on a specified line or line segment. The least squares variant of this model will be discussed in some detail in Chapter 5.

3.4.2   Two or More Medians

As a further generalisation of the basic model, we may replace the single ring by a pair of rings and connect some of the observed points to one of the rings whilst the remaining points are connected to the second ring. In this way we will determine a pair of mediancentres which minimise the potential energies associated with each of the two subpopulations. If we wish to determine an optimal partition of the original observations into two groups, then, in principle, we have to determine the optimal position of the pair of rings for every possible allocation of the observations, and to choose the combination which minimises the sum of the two distance functions associated with this choice.

This mechanical model forms the basis of the least sum of absolute distances clustering of a set of observations into two groups. Other forms of clustering are associated with the mechanical models discussed in Chapters 4 and 6.

3.4.3   Transportation Network with Multiple Nodes

In Subsection 3.4.2 we supposed that each of a set of observations were connected to one of two rings. We now suppose that the two rings are also connected to each other. This problem arises naturally in network theory when we are concerned with the problem of identifying a system of roads which connect a given set of locations with the shortest total path length. For example, suppose that we

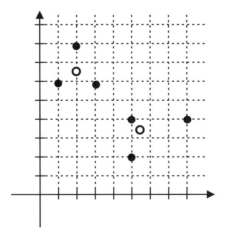

Figure 3.20   Least sum of absolute distances clustering of a set of observations.

are given four towns situated at the corners, A, B, C, and D of a convex quadrilateral, and that we wish to join these four towns by a system of five roads of shortest possible length. The four corners of the quadrilateral would normally be represented by holes drilled through a horizontal board, but in the present context it is convenient to dispense with most, if not all, of the board, and to represent these four points by eyelets located at appropriate positions in a horizontal plane. Two rings are placed at arbitrary positions in the interior of this quadrilateral. Lengths of string with unit weights attached to their lower ends are passed through two adjacent eyelets and joined to one of the rings. In a similar way, two strings with unit weights at their lower ends are passed through the other two eyelets and joined to the second ring. Finally, a length of string with a unit weight at its lower end is passed through the second ring and tied to the first. In principle, this mechanical model will not collapse as the rings are constrained to lie in the horizontal plane defined by the four eyelets, see Figure 3.21a. However, in practice, we would have to attach a balloon filled with sufficient gas to the second ring in order to support the unit weight suspended from it, see Figure 3.21b. The vertical position of this balloon does not change and thus it does not enter into the potential energy calculations.

Each of the strings in this model is under unit tension. Thus, if we treat one of the interior rings as if it were fixed, then it is clear from our earlier results that the position of the other ring is deter-

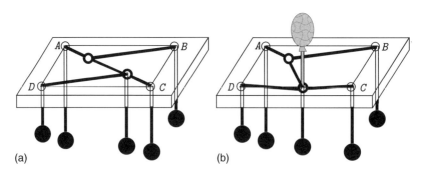

Figure 3.21 (a) Formal mechanical model, and (b) physical model of a simple transportation network problem.

mined in such a way that the three strings connected to it subtend an angle of 120 degrees with each other. Further, the same situation must hold at the second ring. This characterisation of the solution to the network problem is usually associated with the name of Jakob Steiner.

The result given here is restricted to the case in which we have decided to incorporate two additional rings in the model and to connect them to the four given points in a particular way. A much more general result is given in Section 3.5. In this more general solution we no longer need to specify the number of interior nodes, nor do we need to specify how these nodes are to be connected to the given points. As is to be expected, the formal mathematical theory underlying this result is well beyond the scope of this book. Nevertheless, it is still possible to outline a satisfactory mechanical model for the solution of this problem, as we shall now see.

3.5   Mechanical Models in Sums of Areas

3.5.1   Soap Bubble Models

As a natural generalisation of the class of problems discussed so far in this chapter, we consider those which minimise a sum of areas instead of a sum of lengths. In its most general form, this is a species of problem in the calculus of variations. The optimality criterion may again be interpreted as a potential energy function,

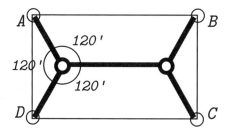

Figure 3.22   Solution of the simple transportation network problem.

but now the physical mechanism used for minimising the surface are of a given geometrical figure subject to certain constraints is that of surface tension in an elastic film such as a soap bubble. In this context, it should be noted that the spherical shape of a freely floating soap bubble may be taken as indicating that the sphere is the three-dimensional figure with minimal surface area for a given volume.

The particular formula recommended by Courant (1940) and Courant and Robbins (1941) for a solution which produces a long-lasting elastic film is obtained by dissolving 10 grams of pure dry sodium oleate in 500 millilitres of distilled water before mixing 15 cubic units of this solution with 11 cubic units of glycerin. They found that films produced by this solution are robust to deformation and sometimes last for several minutes.

Suppose that we wish to determine the system of plane or curved surfaces which minimise the surface area of a figure whose boundaries are given by the edges of a closed three-dimensional rib structure. Courant (1940) and Courant and Robbins (1941, pp. 387–395), following Joséph Antoine Ferdinand Plateau's work of 1873, discuss the experimental solution of problems of this type in some considerable detail. A wire figure of the given shape is immersed in a bath of liquid with low surface tension. It is then carefully withdrawn to reveal a physical determination of the optimal solution to the corresponding geometrical problem.

For example, if a wire figure outlining the boundaries of a cube is immersed in the bath and withdrawn with sufficient care, then a figure with thirteen surfaces will emerge. At first sight, it might be thought that the figure with minimum surface area should consist of twelve isosceles triangles with a common apex at the centre of the cube and bases on its twelve edges. In fact, a lesser surface area is obtained by replacing the central point by a small square lying intermediately between, and parallel to, a pair of opposite sides. This central square is connected to the twelve edges by eight trapezia and four isosceles triangles, see Figure 3.23.

Nor is this the end of the matter, as a further reduction in potential energy is obtained by replacing the plane surfaces of this geometrical figure by the curved surfaces indicated in Figure 240 of Courant and Robbins (1941, p. 387).

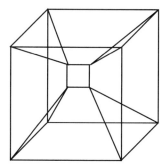

Figure 3.23 Thirteen-faced figure whose boundary is a cube.

Meihle (1958), following Courant (1940), has used a similar procedure to obtain a physical solution to the transportation network problem of Subsection 3.4.3. The n locations of the towns are marked at matching points on a pair of transparent horizontal planes. The pairs of marked points are then connected by vertical rods of unit length. The resulting rigid structure is immersed in a bath of liquid and carefully withdrawn to reveal the solution of the problem. In the special case when there are four rods located at the corners of a convex quadrilateral, the solution consists of five vertical planes set at 120 degrees to each other as indicated in Figure 3.24.

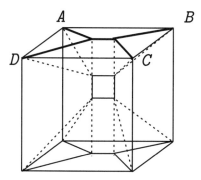

Figure 3.24 Alternative solution of the simple transportation network problem.

By contrast with the result given in Subsection 3.4.3, it should be noted that the present solution is obtained without restriction on the number of interior nodes and without restriction on the manner in which these nodes are connected to each other, or to the n given points.

3.5.2   Oja's Spatial Median

A related problem is that of Oja's spatial median. Suppose that we are given a set of n observations on two variables. Then we may represent these n observations as a scatter of n points in the two-dimensional xy-plane. An additional point is marked at an arbitrary position in the same plane. Each pair of given points taken together with this additional point forms a triangle. There are thus a total of $^nC_2$ distinct triangles of this type. The areas of these $^nC_2$ triangles are summed and the position of the additional point chosen in such a way as to minimise the total area. The optimal position of the additional point defines Oja's (1983) spatial median.

Figure 3.25 illustrates the particular case when there are four points, A, B, C, and D in the form of a convex quadrilateral. The additional point P is assumed to lie in the interior of this quadrilateral, and is connected to the sides AB, BC, CD, and DA in such a way as to form four triangles ABP, BCP, CDP, and DAP, whose total area is equal to that of the quadrilateral $ABCD$. The point P is also connected to the diagonals AC and BD in such a

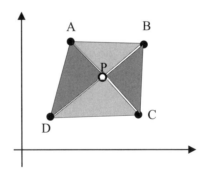

Figure 3.25   Mechanical determination of Oja's spatial median.

way as to form two further triangles *ADP* and *ACP*. The area of these additional triangles, and thus the total area of all six triangles, is clearly minimised by placing the point *P* at the intersection of the diagonals *AB* and *BC*. Thus, in this context, the Oja spatial median is uniquely defined by the point of intersection of the diagonals of the given convex quadrilateral.

In general, this problem is clearly not amenable to the approach discussed in this section as the plane areas of the individual triangles will usually overlap. However, in practice, the problem is solved by rewriting it in the form of a linear programming problem and using the standard simplex procedure to solve it, see Nyblom, Niinimaa, and Oja (1992) for details.

3.5.3   Functional Approximation

The minimum area technique discussed in this section may readily be extended to the problem of approximating a given curve by a sequence of one or more line segments as illustrated in Figure 3.26. However, this simple generalisation does not seem to offer any insight into the essential nature of the fitting problem beyond that indicated by the diagram alone, and we shall not discuss this technique further in this book. For a detailed exposition of fitting procedures of this type, see Watson (1980).

The concept of the area between an observed and a fitted curve may also be generalised to powers of the distance between

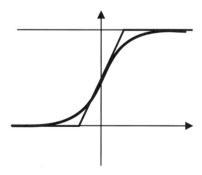

Figure 3.26   Approximating a curve by a sequence of line segments

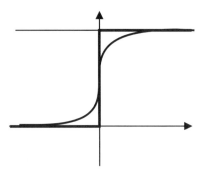

Figure 3.27   Approximating a step function by a curve.

these curves integrated over a suitable range of values of the defining parameter. In particular, we may choose to minimise the maximum difference between the observed and the fitted curve. Again, for a detailed exposition of fitting procedures of this type, see Watson (1980).

As well as forming the basis of a fitting procedure, the maximum difference between an observed cumulative distribution function and a hypothetical curve of the type illustrated in Figure 3.27 may be used as the basis of a Kolmogorov-Smirnov test of the hypothesis that the underlying distribution takes a particular statistical form.

References

Courant, R. (1940), Soap film experiments with minimal surfaces, *American Mathematical Monthly* **47**: 167–174.

Courant, R. and H. Robbins (1941), *What is Mathematics?*, Oxford University Press, New York.

Franksen, O. I. and I. Grattan-Guinness (1989), The earliest contribution to location theory? Spatio-economic equilibrium with Lamé and Clapeyron 1829, *Mathematics and Computers in Simulation* **31**: 195–220.

Gower, J. C. (1974), Mediancentre, *Applied Statistics* **23**: 466–470.

Kuhn, H. W. (1967), On a pair of dual nonlinear programs, in J. Abadie (Ed.), *Methods of Nonlinear Programming*, North-Holland Publishing Company, Amsterdam, 29–54.

Meihle, W. (1958), Link-length minimisation in networks, *Operations Research* **6**: 232–243.

Niinimaa, A., H. Oja and J. Nyblom, (1992), The Oja Bivariate Median, *Applied Statistics*, **41**: 611–617.

Oja, H. (1983), Descriptive statistics for multivariate distributions, *Statistics and Probability Letters* **1**: 327–332.

Vinod, H. D. (1969), Integer programming and the theory of grouping, *Journal of the American Statistical Association* **64**: 516–519.

Watson, G. A. (1980), *Approximation Theory and Numerical Methods*, John Wiley and Sons, London, England.

CHAPTER 4

Method of Least Squared Deviations

4.1 One- and Two-Dimensional Means

4.1.1 Mechanical Models for One-Dimensional Means

At this point, it is convenient to interrupt our discussion of the least sum of absolute deviations problem to consider the more familiar least sum of squared deviations problem. As in Chapter 3, we begin by considering the problem of fitting a single parameter to a set of n equally reliable observations on a single variable. Further, since we may do so without adding much to the complexity of our discussion, we consider the one-dimensional problem as a special case of the corresponding two-dimensional problem. We therefore assume that the n observations on the variables X and Y are represented by a set of n points in the two-dimensional Cartesian plane, and that these points are marked on a horizontal board relative to a given y-axis that runs from south to north, and a given x-axis that runs from west to east in the horizontal plane. We drill a hole through the horizontal board at each of the observed points and mark the corresponding point on a second horizontal board at unit distance below the first. For each observation, we attach a spring of unit natural length and unit modulus to the marked point on the lower board, pass it through the corresponding hole in the upper board, and attach its upper end to a ring at an arbitrary point in the upper horizontal plane. If this ring is placed on one of the holes in the upper board, then the corresponding spring will have unit length. Further, if the ring is moved a certain distance away from this hole, then the spring will be extended beyond its natural length by the same amount and the potential energy in the stretched spring will be proportional to the square of its extension. By summing the potential energy functions for the n springs we obtain the corresponding function of the system as a whole.

To simplify our discussion of mechanical models based on springs, we shall henceforth abstract from reality and assume that the springs employed in our theoretical models are *ideal springs* with zero natural lengths. We are thus able to dispense with the second horizontal board which features in our model as a device for absorbing the nonzero natural lengths of real springs. In this revised model, there is a single horizontal board, and the springs are attached to the marked points on this board.

We are now able to consider the case of a set of n equally reliable observations $y = y_1, y = y_2, \ldots, y = y_n$ on a single unknown quantity Y. We may represent these n observations by a set of n points lying along a straight line which runs from south to north in the horizontal plane. If the current position of the ring is at the point $y = a$, then the ith observation $y = y_i$ is at a distance $|y_i - a|$ from the ring and contributes a quantity proportional to $P_i = \frac{1}{2}(y_i - a)^2$ to the overall potential energy function

$$P = \frac{1}{2}\sum(y_i - a)^2$$

Our problem in this section is to choose a value for the parameter a in such a way as to minimise this scaled sum of squared deviations function. Now, the first derivative of this function with respect to the unknown parameter a

$$\frac{dP}{da} = -\sum(y_i - a)$$

takes a zero value when a is set equal to the arithmetic mean of the observations $\bar{y} = \sum y_i/n$. Further, the second derivative of this function

$$\frac{d^2P}{da^2} = n$$

is positive for all values of a, and the potential energy function is thus minimised when a is set equal to $\bar{y}$.

We may illustrate the solution of this problem by plotting the individual potential energy functions for each of the n observations in the vertical plane with the parameter a on the horizontal axis and P_i on the vertical axis, see Figure 4.1. These individual functions are summed to obtain the overall potential energy function P that is to be minimised.

However, by contrast with the sum of absolute deviations problem discussed in Chapter 3, the insight gained from this figure is slight in the one-dimensional case and nugatory in higher dimensions. We therefore abandon the geometrical approach to this problem in favour of a mechanical one.

4.1.2 Balance of Forces in the One-Dimensional Case

The discussion in Subsection 3.2.3 suggests an alternative approach to the problem of Subsection 4.1.1 based on the balance of forces in the springs of the mechanical model. In this context, we find that the force exerted on the ring by the ith spring is proportional to the extension of this spring and acts in the direction of the ith observation. Thus, a spring anchored at a point $y = y_i$ will exert a force proportional to $y_i - a$ acting in a northwards direction or, equivalently, to a force proportional to $a - y_i$ acting in a southwards direction. Thus, the net force acting on the ring is given by $\sum(y_i - a)$ acting in a northwards direction and this force function may again be identified with the negative of the derivative of the potential energy function

$$\frac{dP}{da} = -\sum(y_i - a)$$

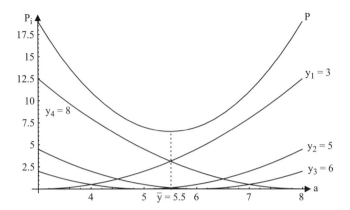

Figure 4.1 Parabolic potential energy functions and their sum.

The ring is clearly in a state of equilibrium when this net force takes a zero value, and we may deduce that the arithmetic mean of the observations $\bar{y}$ is the appropriate value of a in the one-dimensional case.

4.1.3 Mechanical Model in the Two-Dimensional Case

Generalising this mechanical model to the two-dimensional case, we find that we have to attach a spring of unit modulus running from the ith given point (x_i, y_i) to a ring at the arbitrary point (c, a). The distance between these two points is

$$\sqrt{[(x_i - c)^2 + (y_i - a)^2]}$$

so that this spring contributes an amount proportional to the square of this distance

$$P_i = \frac{1}{2}[(x_i - c)^2 + (y_i - a)^2]$$

to the overall potential energy function

$$P = \frac{1}{2}\sum[(x_i - c)^2 + (y_i - a)^2]$$

Our problem is again to choose values for the parameters c and a in such a way as to minimise this scaled sum of squared deviations function. Now, the first order partial derivatives of this function with respect to c and a

$$\frac{\partial P}{\partial c} = -\sum (x_i - c)$$

and

$$\frac{\partial P}{\partial c} = -\sum (y_i - a)$$

clearly take zero values when c is set equal to the arithmetic mean of the x-observations $\bar{x} = \sum x_i/n$ and a is set equal to the arithmetic mean of the y-observations $\bar{y} = \sum y_i/n$. Further, the same-parameter second order partial derivatives of this function

$$\frac{\partial^2 P}{\partial c^2} = n$$

and

$$\frac{\partial^2 P}{\partial a^2} = n$$

are positive for all values of c and a, and the cross-parameter second order partial derivative

$$\frac{\partial^2 P}{\partial c \partial a} = 0$$

is zero, so that the second order conditions for a minimum are satisfied and the potential energy function P is minimised when c is set equal to $\bar{x}$ and a is set equal to $\bar{y}$.

4.1.4 Balance of Forces in the Two-Dimensional Case

Alternatively, we may obtain the same results by considering the balance of forces in the springs of the model. In the present context, the ith spring exerts a force on the ring which is proportional to the extension in this spring

$$\sqrt{[(x_i - c)^2 + (y_i - a)^2]}$$

and directed in such a way as to pull the ring towards the point (x_i, y_i). Applying Pythagoras's theorem to this force, we find that it may be resolved into a force proportional to $x_i - c$ pulling the ring eastwards and a force proportional to $y_i - a$ pulling it northwards. Summing these individual contributions, we find that the system of springs pulls the ring eastwards with a force proportional to the sum of the deviations in the x-direction $\sum(x_i - c)$, and northwards with a force proportional to the sum of deviations in the y-direction $\sum(y_i - a)$. Comparing these expressions with those of the previous section, we find that the negatives of these resolved force functions may again be identified with the partial derivatives of the potential energy function P. Further, setting each of these expressions equal to zero, we find that the ring is in a state of equilibrium when c is set equal to the arithmetic mean of the x observations $\bar{x} = \sum x_i/n$ and a is set equal to the arithmetic mean of the observations $\bar{y} = \sum y_i/n$. The point $(\bar{x}, \bar{y})$ is known as the centre of gravity or *centroid* of the set of observations.

Finally, we note that the potential energy function discussed in this section may be written as the sum of two functions, one expressed in terms of c and the other in terms of a. We may therefore treat the determination of a and c as if they were algebraically independent of one another.

4.1.5 The Modulus of the Parameter Estimates

In the preceding analysis, we have implicitly assumed that the n observations on Y are distinct. We shall now assume that these n observations take one of r distinct values. If there are m_i observations associated with the value $y = y_i$, then there are m_i springs of unit modulus associated with this value and contributing m_i terms of the form $\frac{1}{2}(y_i - a)^2$ to the overall potential energy function. Instead of representing these m_i observations by m_i distinct springs of unit modulus, we may substitute a single spring of modulus m_i with the same effect on the potential energy function. Thus, the contribution to the overall potential energy function associated with the spring of modulus m_i is proportional to the function

$$P_i = \frac{1}{2} m_i (y_i - a)^2$$

We have already established that the first derivative of this function

$$\frac{dPi}{da} = -m_i (y_i - a)$$

is the negative of the force pulling the ring in the y-direction. Similarly, the second derivative of this function

$$\frac{d^2 Pi}{da^2} = m_i$$

clearly defines the modulus of the ith spring.

Having explained how to recover the modulus of a single observation from the corresponding potential energy function, we may now use the same technique to determine the moduli of the parameter estimates. Applying these results in the context of the potential energy function associated with a set of r springs of modulus $m_i, m_2, \ldots, m_r$, we find that the potential energy function of the system is proportional to

$$P = \frac{1}{2} \sum m_i (y_i - a)^2$$

The first derivative of this function

$$\frac{dP}{da} = -\sum m_i (y_i - a)$$

is again the negative of the force acting on the ring. It takes a zero value when the potential energy function is minimised, and this occurs when a is set equal to the weighted arithmetic mean of the observations

$$\bar{y}_m = \frac{\sum m_i y_i}{\sum m_i}$$

Further, the second derivative of the potential energy function

$$\frac{d^2P}{da^2} = \sum m_i$$

gives the modulus of this estimate $\bar{y}$ of a.

This result is most familiar in the case of a set of n equally reliable observations, that is, when $r = n$ and $m_1 = m_2 \ldots = m_n = 1$. In this context, a modulus of $\sum m_i = n$ for unweighted arithmetic mean of the observations $\bar{y} = \sum y_i/n$ corresponds to the familiar expression σ^2/n for the variance of this mean. Note that the positive scalar σ^2 may be supposed to absorb the unknown factor of scale implicit in the definition of the potential energy function.

This technique for recovering the moduli of the parameter estimates from the potential energy function associated with a set of observations may readily be extended to the two-dimensional case. This mechanical interpretation of the sum of squared deviations fitting problem was first expounded in a substantial monograph by William Fishburn Donkin in 1844.

Finally, we note that the functions

$$\bar{y}_m = \frac{\sum m_i y_i}{\sum m_i} \quad \text{and} \quad 2P = \sum m_i(y_i - \bar{y}_m)^2$$

are usually referred to as the weighted arithmetic mean and the weighted sum of squared deviations functions respectively. This traditional terminology is unfortunate in the present context as we should have liked to reserve the term "weight" for the corresponding concept in the least sum of absolute deviations problem. However, this usage derives from the work of Carl Friedrich Gauss and Pierre Simon Laplace in the early years of the nineteenth century, and it is now clearly far too late to attempt to effect a change.

4.2 Linear, Curvilinear, and Regional Constraints

The various forms of constrained estimation problems discussed in Chapter 3 may readily be extended to the least squares case. As in Section 3.3, we may insist that the fitted point must lie on a given line or in a given region. The mechanical models involved are exactly the same except that the strings under unit tension are replaced by springs of unit modulus. The diagrams are essentially the same except that there are now no simple rules for determining the location of the optimal point. We shall therefore not trouble to repeat this material in the present context.

Instead, we address a special case in which a ring is attached by springs to two given points in the horizontal plane, but is constrained to lie on the circumference of a circular disc in the same plane. This problem is particularly interesting as it is possible for the ring to have a stable point of equilibrium on either side of the fixed circle.

In Section 3.3, we briefly discussed the determination of the optimal position of a ring which is pulled towards two points in the plane by strings under unit tension, but is constrained to lie on the circumference of a circular disc in the same plane. If one of the points is to the north of the circle whilst the other is to the south, then in the absence of any constraint, any point on the line segment joining these two points will minimise the sum of the distances from the ring to the two points. Thus, when the circular constraint is imposed, we find that either of the points on the intersection of the circle with this line segment will serve as an optimal solution to the constrained least sum of absolute deviations problem.

With a view to clarifying our discussion of the corresponding least sum of squared deviations problem, we re-examine this problem in a slightly different form. We treat the second fixed point as if it were movable and start with it close to the northern point. We move it round the eastern side of the circle until it arrives at its final position to the south of the circle. Initially, the line joining the two points will not intersect the fixed circle and there will be a unique point on the circle representing a minimum of the sum of the distances to the two points, see Figure 4.2.

Then, as the second point is moved round the eastern side of the circle, we reach a certain stage at which the line segment joining the two points is tangential to the fixed circle, see Figure 4.3.

Subsequently, this line segment will cut the fixed circle in two points and either of these points of intersection will represent a minimum of the sum of distances function. When the second point

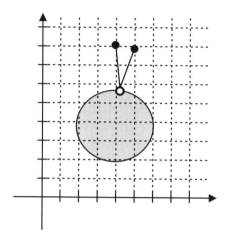

Figure 4.2   Circular constraint problem – initial position.

approaches its final position to the south of the fixed circle, there will be two optimal positions for the ring; one at the northernmost point of the circle and one at its southernmost point, see Figure 4.4.

Further, if instead of passing the movable point round the eastern side of the circle, we pass it round the western side, then

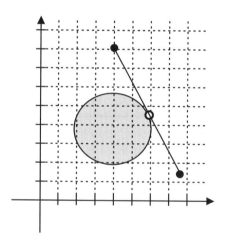

Figure 4.3   Circular constraint problem – tangential position.

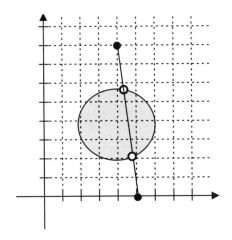

Figure 4.4   Circular constraint problem – intermediate position.

we obtain a parallel set of results which again confirm the same two points as the optimal solutions to the original problem, see Figure 4.5.

Now, we adjust this technique to the problem of choosing a point on a fixed circle in such a way as to minimise the sum of the

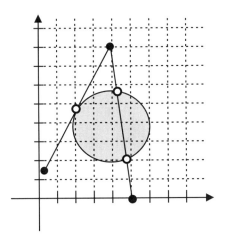

Figure 4.5   Circular constraint problem – alternative approaches to the final position.

squared distances from the given points to a ring at an arbitrary point on the circumference of the circle. Again, we place the second (movable) point near the first to the north of the fixed circle, and gradually move it round the eastern side of the circle until it reaches its final position to the south of the circle. We thus obtain a sequence of diagrams of the type illustrated in Figures 4.2, 4.6, and the right side of Figure 4.7.

By contrast with the case discussed above, we find that each stage of this procedure determines a single point on the eastern side of the circle; hence one of the solutions to the sum of squared deviations problem must lie on this side of the circle. However, we may also pass the movable point round the western side of the fixed circle and obtain a second solution on that side. Thus, we again have two solutions to the fitting problem; but, in the present context, one of the solutions lies on the eastern side of the circle and one on the western side, see Figure 4.7.

Clearly this problem defines an interesting identification problem that deserves further attention. In fact, it can be shown that there is a small region (not shown and possibly empty) to the south of the fixed circle in Figure 4.7. If the second fixed point lies outside this region, then, as illustrated in Figures 4.2 and 4.6, there is a single stable solution to the problem; and if it lies within this region then, as illustrated in Figure 4.7, there are two stable solutions. Further, as the movable point enters the region from the west, there will be no obvious change in the nature of the solution, but as it passes through the eastern boundary there will be a jump discontinuity as the ring passes rapidly round the southern portion of the fixed circle to reach the stable solution on its eastern side. Similarly, as the second point passes from east to west through this region a jump discontinuity will occur when the movable point reaches its western boundary.

This mechanical model was developed by Christopher Zeeman to illustrate the workings of the so-called "cusp catastrophe" of catastrophe theory. For a detailed description of a second instance of this type of catastrophe in which a soap film alternates between a two-sided figure and a one-sided Möbius strip as a circular frame is continuously deformed, see Courant (1940, p. 169) or Courant and Robbins (1941, pp. 388–389).

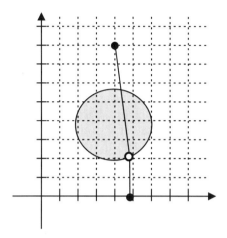

Figure 4.6   Circular constraint problem – least squares intermediate position.

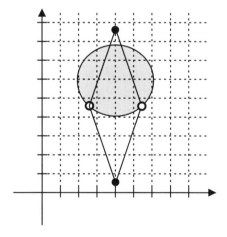

Figure 4.7   Circular constraint problem – least squares final positions.

4.3 Simple Linear Regression

4.3.1 Orientation

The problem of fitting a single point of best fit to a set of n points in the xy-plane may readily be generalised to the problem of fitting a line of best fit to the same set of n points. However, by contrast with the problem of Section 4.1, in the case of the line fitting problem we have to choose an orientation in which distances are to be measured. This difficulty does not arise in the simpler case as the distance of one point from another is explicit, whereas when we are interested in measuring the distance from a point to a line, the measure will depend on which point on the line is chosen as the basis of the definition.

In this chapter, as in Chapter 1, we shall discuss two traditional orientations. The first measures distances parallel to the y-axis whilst the second measures distances perpendicular to the fitted line.

Plotting the n given observations on the variables X and Y as points in the horizontal xy-plane, we insert an arbitrary line in this diagram and connect each of the points to it by a line segment drawn parallel to the y-axis. Applying the criterion of this chapter, we have to choose the position of this line in such a way as to minimise the sum of the squared distances from the given points to the fitted line, see Figure 4.8.

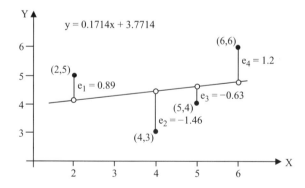

Figure 4.8 Conventional line fitting problem.

If the ith point is located at (x_i, y_i) and the arbitrary line is defined by the equation $y = a + bx$ then the point $(x_i, a + bx_i)$ will lie on the line and the ith given point is at a (positive or negative) distance $e_i = y_i - a - bx_i$ from this line when distances are measured parallel to the y-axis. In this context, we have to choose values for the parameters a and b in such a way as to minimise the sum of the squared distances $\sum e_i^2$. The solution of this fitting problem is illustrated in Figure 4.8 for the data given in Table 2.1.

For the alternative orthogonal line fitting procedure, we again plot an arbitrary line in the scatter of points in the xy-plane but, this time, we join the points to the fitted line by line segments that are drawn perpendicular to the fitted line, see Figure 4.9.

Now, the arbitrary line $y = a + bx$ has a slope of b, so that the perpendicular to this line must have a slope of $-1/b$ and reference to Figure 4.9 shows that the distance $e_i = y_i - a - bx_i$ measured parallel to the y-axis may be resolved into a distance of $e_i/\sqrt{(1 + b^2)}$ measured perpendicular to the given line and a distance of $be_i/\sqrt{(1 + b^2)}$ measured parallel to this line. Again, applying the criterion of this chapter, we have to choose values for the parameters of the fitted line in such a way as to minimise the sum of the squared distances from the given points to this line. That is, we have to choose values for a and b in such a way as to minimise the scaled sum of squared deviations function $\sum e_i^2/(1 + b^2)$.

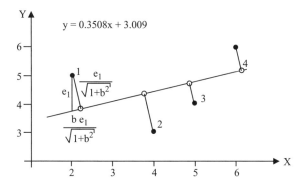

Figure 4.9 Orthogonal line fitting problem.

4.3.2 Mechanical Model and Balance of Forces in the Conventional Case

Building on the results in Section 4.1, we may express either of these fitting problems in a mechanical form by attaching one end of a spring of unit modulus and zero natural length to each of the given points and the other end to a ring which is passed over a rigid rod representing the fitted line. In the orthogonal case the springs are free from constraints and, in equilibrium, they will lie perpendicular to the fitted line. By contrast, in the conventional case, the springs are constrained to lie parallel to the y-axis. In this context, the fitted lines in Figures 4.8 and 4.9 are to be interpreted as rigid rods and the line segments connecting the n observations to these lines as the current positions of the springs.

By the argument of Section 4.1, it is readily established that the potential energy of the system of springs is given by the function

$$P = \frac{1}{2} \sum e_i^2$$

in the conventional case. The force exerted by a stretched spring is proportional to its extension, and, in the present context of ideal springs, this force is proportional to the length of the spring.

In the conventional orientation, it is convenient to describe the y-axis as running from south to north and the x-axis as running from west to east. Each of the given points lying to the north of the fitted line will exert a force on the line tending to pull it northwards whilst those lying to the south will tend to pull it southwards. In equilibrium, these two sets of forces must balance. Similarly, each of the springs will exert a rotational force or couple on the rod. Again, when the rod is in equilibrium, the couples tending to turn the rod in a clockwise direction about an arbitrary point on the rod must balance those tending to turn it in an anticlockwise direction.

Suppose that the fitted line has intercept parameter a and slope parameter b, then its equation may be expressed in the form $y = a + bx$ and the point $(x_i, a + bx_i)$ on this line will lie due north or due south of the point (x_i, y_i) representing the ith observation. Thus, the ith observation is at a distance $e_i = y_i - a - bx_i$ due north or due south of the fitted line. Note that e_i takes a positive value if the ith observation is to the north of the line and a negative value if it is to the south. The net northwards force in the ith spring is given by e_i. For the forces in the springs to balance, we must have $\sum e_i = 0$. Similarly, the ith spring exerts a couple about the point $(0, a)$ at the intersection of the line with the y-axis. The value of this couple is obtained by multiplying the force in the ith spring e_i by the perpendicular distance (x_i) between the ith point and the y-axis. Thus, we must have $\sum x_i e_i = 0$ if these couples are to balance.

This mechanical representation of the conventional least squares problem was developed in an abstract physical form by Simon Newcomb in 1873. An alternative parameter space model for the same procedure will be outlined in Section 5.3.2.

Most readers of this book will recognise the algebraic expressions $\sum e_i = 0$ and $\sum x_i e_i = 0$ as the first order conditions for the minimisation of the sum of the squared deviations function $2P = \sum e_i^2$. They will also know that, for these conditions to be satisfied, we have to set $b = \hat{b}$ and $a = \bar{y} - \bar{x}\hat{b}$ where

$$\hat{b} = \frac{\sum(x_i - \bar{x})(y_i - \bar{y})}{\sum(x_i - \bar{x})^2}$$

The corresponding minimum value of the sum of squares function being given by $2P^*$ where

$$P^* = \frac{1}{2}\sum(y_i - \bar{y})^2 - \frac{1}{2}[\sum(x_i - \bar{x})(y_i - \bar{y})]^2 / \sum(x_i - \bar{x})^2$$

In this context, it is clear that the fitting-by-eye procedure of Section 2.5 may be related to the conventional least squares procedure by explaining the mechanical foundations of the latter procedure and supplying the sum of forces and sum of couples functions as indicators of the difference between the present position of the fitted line and that of the least squares line. Similar adjustments of the proposed computer package may be used to relate the fitting-by-eye procedure to the orthogonal least squares procedure discussed below, and to the conventional and orthogonal variants of the least sum of absolute deviations and L_p-norm procedures of Chapter 5.

4.3.3   Mechanical Model in the Orthogonal Case

By the argument of Subsection 4.3.1, it is readily established that the potential energy function associated with the orthogonal least squares line fitting problem is given by

$$P = \frac{1}{2} \frac{\sum e_i^2}{(1 + b^2)}$$

The first order partial derivatives of this function with respect to the parameters a and b are given by

$$\frac{\partial P}{\partial a} = -\sum e_i / (1 + b^2)$$

and

$$\frac{\partial P}{\partial b} = \frac{-\sum x_i e_i}{(1 + b^2)} - \frac{\sum b e_i^2}{(1 + b^2)^2}$$

Setting these partial derivatives equal to zero, we have from the first equation, $a = \bar{y} - b\bar{x}$, and, on substituting this expression into the second equation, we find that we have to solve what seems to be a cubic equation in b

$$(1 + b^2)(s_{xy} - b s_{xx}) + b(s_{yy} - 2b s_{xy} + b^2 s_{xx}) = 0$$

but which simplifies to the quadratic equation

$$s_{xy} + b(s_{yy} - s_{xx}) - b^2 s_{xy} = 0$$

where $s_{xx} = \sum (x_i - \bar{x})^2$, $s_{xy} = \sum (x_i - \bar{x})(y_i - \bar{y})$, and $s_{yy} = \sum (y_i - \bar{y})^2$.

This equation has two solutions

$$b = \frac{(s_{yy} - s_{xx}) \pm \sqrt{[(s_{yy} - s_{xx})^2 + 4 s_{xy}^2]}}{2 s_{xy}}$$

one of which corresponds to the maximum value of the potential energy function P, and the other to its minimum value. In fact, we can show that the $+$ form yields the smaller value of $m = 2P$ as

$$m = \frac{1}{2}(s_{xx} + s_{yy}) \mp \sqrt{[(s_{yy} - s_{xx})^2 + 4s_{xy}{}^2]}$$

An alternative realisation of this solution is obtained by writing the partially optimised potential energy function in the form

$$2P = \frac{(s_{yy} - 2bs_{xy} + b^2 s_{xx})}{(1 + b^2)}$$

and then setting $b = -u/v$ to obtain the ratio of quadratic forms

$$2P = \frac{(v^2 s_{yy} + 2uv s_{xy} + u^2 s_{xx})}{(u^2 + v^2)}$$

Clearly, the maximum and minimum values of P are given by the maximum and minimum eigenvalues of the 2×2 matrix

$$\begin{bmatrix} S_{xx} & S_{xy} \\ S_{xy} & S_{yy} \end{bmatrix}$$

whilst the required values of b may be derived from the scaled eigenvectors corresponding to these eigenvalues.

Readers who are not familiar with matrix algebra may omit the remainder of this section as this material is included for the convenience of those who are.

Denoting the typical eigenvalue of this matrix by m, we have to choose a value for m in such a way that the determinant

$$\begin{vmatrix} S_{xx} - m & S_{xy} \\ S_{xy} & S_{yy} - m \end{vmatrix}$$

is zero. That is, we have to solve the equation

$$m^2 - (s_{xx} + s_{yy})m + s_{xx}s_{yy} - s_{xy}{}^2 = 0$$

Once again, the solutions of this quadratic equation are given by

$$m = \frac{1}{2}[s_{xx} + s_{yy} \mp \sqrt{[(s_{yy} - s_{xx})^2 + 4s_{xy}{}^2]}$$

Substituting these values in the matrix equation

$$\begin{bmatrix} s_{xx} - m & s_{xy} \\ s_{xy} & s_{yy} - m \end{bmatrix} \begin{bmatrix} b \\ -1 \end{bmatrix} = \begin{bmatrix} 0 \\ 0 \end{bmatrix}$$

we find that the values of $b = -v/u$ defining the eigenvectors corresponding to these eigenvalues are given by

$$b = (s_{yy} - m)/s_{xy}$$
$$= \frac{s_{yy} - s_{xx} \pm \sqrt{[(s_{yy} - s_{xx})^2 + 4s_{xy}^2]}}{2s_{xy}}$$

as before. In passing, we note that this solution derives from the matrix formulation of the principal components problem of Chapter 11.

4.3.4 Balance of Forces in the Orthogonal Case

In the orthogonal case, and for a given position of the arbitrary line, we may partition the observations into those tending to pull the line in one direction and those tending to pull it in the opposite direction. These two sets of forces must balance when the fitted rod is in equilibrium. Further, by selecting an arbitrary point on the fitted line, say at its intersection with the y-axis, we find that the turning forces or couples exerted by these two sets of springs must also balance when the line is in equilibrium.

Again supposing that the fitted line has intercept parameter a and slope parameter b, then its equation may be expressed in the form $y = a + bx$ and the point $(x_i, a + bx_i)$ will lie due north or due south of the point (x_i, y_i). Thus, the ith observation is at a distance $e_i = y_i - a - bx_i$ due north or due south of the fitted line and at a distance $e_i/\sqrt{(1 + b^2)}$ when such distances are measured perpendicular to the fitted line. Thus, the net force in the ith spring is proportional to its extension $e_i/\sqrt{(1 + b^2)}$. And, for the forces in the springs to balance, we must have

$$\sum e_i/\sqrt{(1 + b^2)} = 0$$

Similarly, the distance from $(0, a)$ to the point $(x_i, a + bx_i)$ is x_i when measured parallel to the x-axis and $x_i\sqrt{(1 + b^2)}$ when measured parallel to the fitted line. However, we are concerned with the distance from $(0, a)$ to (x_i, y_i) measured parallel to the fitted line, so that this figure must be increased by a (positive or negative) amount $be_i/\sqrt{(1 + b^2)}$ before being multiplied by the force in the ith spring $e_i/\sqrt{(1 + b^2)}$ to yield an expression for the ith couple

$$x_i e_i + be_i^2/(1 + b^2)$$

In equilibrium, these couples must sum to zero

$$\sum x_i e_i + \sum be_i^2/(1 + b^2) = 0$$

These expressions may readily be identified as negative multiples of the partial derivatives given in Subsection 4.3.3. The scale factor in these expressions arises because the derivatives are taken perpendicular to the ba-plane whereas the forces are taken perpendicular to the fitted line. However, we shall not discuss the orthogonal least squares problem in any greater detail in this book as to do so would take us too far out of our way.

4.4 Moduli of Parameter Estimates

4.4.1 Direct Evaluation

Returning to the conventional form of the least squares line fitting problem, but now assuming that the moduli of the springs attached to the points in the horizontal plane take the (generally distinct) values $m_1, m_2, \ldots, m_n$ rather than the common value of unity. In this context, the optimal values of the parameters of the fitted line are chosen in such a way as to minimise the "weighted" sum of the squared deviations potential energy function given by

$$P = \frac{1}{2} \sum m_i(y_i - a - bx_i)^2$$

We find that this function has first order partial derivatives with respect to the parameters a and b

$$\frac{\partial P}{\partial a} = -\sum m_i(y_i - a - bx_i)$$

and

$$\frac{\partial P}{\partial b} = -\sum m_i x_i (y_i - a - bx_i)$$

corresponding to the linear and rotational forces acting on the rigid rod. These forces will take zero values when the system is in a state of equilibrium.

Similarly, the second order partial derivatives of this function

$$\frac{\partial P^2}{\partial a^2} = \sum m_i$$

$$\frac{\partial P^2}{\partial a \partial b} = \sum m_i x_i$$

and

$$\frac{\partial P^2}{\partial b^2} = \sum m_i x_i^2$$

determine the joint moduli of the estimators of the parameters. Some readers will be familiar with these results as the variance matrix of these estimators is proportional to the inverse of the matrix of second order partial derivatives

$$\begin{bmatrix} \sum m_i & \sum m_i x_i \\ \sum m_i x_i & \sum m_i x_i^2 \end{bmatrix}$$

4.4.2   Oblique Transformation

More immediate results on the moduli of the least squares estimators may be obtained by defining the weighted mean $\bar{x}_m = \sum m_i x_i / \sum m_i$, $\bar{y}_m = \sum m_i y_i / \sum m_i$ and the additional parameter $c = a + b\bar{x}_m$. With this new notation, the potential energy function may be written as

$$P = \frac{1}{2}\sum_i m_i [y_i - c - b(x_i - \bar{x}_m)]^2$$

In this form, the function has first order partial derivatives

$$\frac{\partial P}{\partial c} = -\sum m_i[y_i - c - b(x_i - \bar{x}_m)]$$

and

$$\frac{\partial P}{\partial b} = -\sum m_i(x_i - \bar{x}_m)[y_i - c - b(x_i - \bar{x}_m)]$$

These expressions clearly take zero values when c is set equal to $\bar{y}_m$ and b is set equal to

$$b = \frac{\sum m_i(x_i - \bar{x}_m)(y_i - \bar{y}_m)}{\sum m_i(x_i - \bar{x}_m)^2}$$

Similarly, the second order partial derivatives of this function are given by

$$\frac{\partial^2 P}{\partial c^2} = \sum m_i$$

$$\frac{\partial^2 P}{\partial c \partial b} = \sum m_i(x_i - \bar{x}_m)$$

and

$$\frac{\partial^2 P}{\partial b^2} = \sum m_i(x_i - \bar{x}_m)^2$$

Now, $\sum m_i(x_i - \bar{x}_m)$ necessarily takes a zero value, so that the nondiagonal elements of the joint moduli matrix are zero $\partial^2 P/\partial c \partial b = 0$, and we deduce that the moduli of these parameter estimates are $\sum m_i$ and of $\sum m_i(x_i - \bar{x}_m)^2$ respectively. That is, the least squares estimates of a and b given by these expressions have variances of $\sigma^2/\sum m_i$ and $\sigma^2/\sum m_i(x_i - \bar{x}_m)^2$.

In this analysis, the definition of c is adopted without further justification. In fact, it derives from the following Lagrange or LDL' decomposition of the joint modulus matrix of Subsection 4.4.1:

$$\begin{bmatrix} 1 & 0 \\ \bar{x}_m & 1 \end{bmatrix} \begin{bmatrix} \sum m_i & 0 \\ 0 & \sum m_i(x_i - \bar{x}_m)^2 \end{bmatrix} \begin{bmatrix} 1 & \bar{x}_m \\ 0 & 1 \end{bmatrix} = \begin{bmatrix} \sum m_i & \sum m_i x_i \\ \sum m_i x_i & \sum m_i x_i^2 \end{bmatrix}$$

4.4.3 Relative Equilibrium

It would be more convenient to employ a general procedure which does not involve the preliminary identification of a suitable reparameterisation of the potential energy function. Applying the 1844 approach of Donkin to a problem in which the parameter b is required to take a value that is distinct from its optimal value, we find that the force function $\partial P/\partial b$ will have to take a nonzero value in order to keep b at the selected non-optimal value, but all other derivatives of the potential energy function will take zero values. This situation represents a state of relative equilibrium. From the equation $\partial P/\partial a = 0$ we deduce that a is given by $a = \bar{y}_m - b\bar{x}_m$. This expression is substituted into the second equation to yield an expression which does not involve the parameter a

$$\left.\frac{\partial P}{\partial b}\right|_{a=\bar{y}m - b\bar{x}_m} = -\sum m_i x_i [y_i - \bar{y}_m - b(x_i - \bar{x}_m)]$$

from which we may deduce that the optimal value of b is given by

$$b = \frac{\sum m_i x_i (y_i - \bar{y}_m)}{\sum m_i x_i (x_i - \bar{x}_m)} = \frac{\sum m_i (x_i - \bar{x}_m)(y_i - \bar{y}_m)}{\sum m_i (x_i - \bar{x}_m)^2}$$

with modulus

$$\sum m_i x_i (x_i - \bar{x}_m) = \sum m_i (x_i - \bar{x}_m)^2$$

4.5 Influential Observations

4.5.1 Changing Moduli

The least squares results are said to be robust to small changes in the data if small changes in the values of one or more observations or their associated moduli lead to proportionately small changes in the values of the parameter estimates. Conversely, if this is not the case, then they are said to be non-robust. In particular, if the modulus of the spring attached to one of the points is changed by a small amount and leads to a large change in the position of the fitted line, then this observation is clearly influential in determining the position of the line. We may thus identify influential observations or sets of observations by considering these effects.

As an obvious measure of influence, we may divide the change in the value of the slope parameter by the change in the modulus of the spring causing that change. This measure of influence is closely related to the *EIC* measure of influence, and was implicitly proposed by Simon Newcomb in 1873, see Farebrother (1999, p. 168).

4.5.2 Adding Observations

The modulus of a particular spring or set of springs may be increased from zero to one or reduced from one to zero. These changes correspond to the addition or deletion of a set of observations. If a single observation is added to (or deleted from) the data set, then the fitted least squares line will take a new position. Since these two lines will usually intersect, we may speak of the least squares line as being rotated about this point of intersection in such a way as to take up a position which more closely reflects the contribution of the additional observation. If the additional observation is remote from the centroid of the original set, then the fitted line will be substantially deflected from its original position. On the other hand, if the additional observation is close to the centroid of the original data set, then the deflection will be minimal. Observations in the first category are said to be *influential* and those in the second to be less so.

4.5.3 Measures of Influence

As a practical technique for identifying influential observations in any given data set, we may temporarily remove one of the observations from the original set and determine the effect of restoring it. It is also possible to perform the same analysis when removing two or more observations from the original set. This more general technique will identify instances in which none of the observations is influential on its own but may be strongly influential when taken in conjunction with other observations.

4.5.4   Ridge Regression

A possible solution to the problem of influential observations is to introduce additional information. Ideally this additional information should take the form of one or more observations that throw light on the portion of the observation space which was not investigated by the original data set. Unfortunately, this is not likely to be possible in practice. For, if there is an apparent lack of coverage in the given data set, then it is most unlikely to be remedied by seeking new observations. It is implicit in the statement of the *multicollinearity problem* that such additional information is not to be found.

   A practical solution to this problem is to introduce a set of fictitious observations that stabilise the parameter estimates. These fictitious observations are often implicitly defined in such a way that the sum of the squared deviations function is augmented by a positive multiple of the sum of the squared parameter values. In particular, in the two-dimensional case, we have to choose values for a and b to minimise the function

$$\sum (y_i - a - bx_i)^2 + k(a^2 + b^2)$$

for some suitable choice of the positive parameter k. The resulting values are known as ridge regression estimates.

   Other possible solutions to the problem of influential observations are described in Chapters 5 and 7 below. In these chapters, we shall make use of the comparatively robust sum of absolute deviations and median squared deviation (or median absolute deviation) criteria in place of the sum of squared deviations criterion of the present chapter.

4.6 Linear Constraints and Hypothesis Tests

4.6.1 Linear Constraints

In certain circumstances, it is appropriate to impose linear constraints on the values of the parameters. For example, we may assume that the parameters a and b satisfy the condition $a + 2b = 3$ or, more generally, the condition $a + bx_0 = y_0$ where x_0 and y_0 are fixed numbers. This situation may readily be represented in the standard xy-plane diagram of Section 4.3 as this constraint requires that the fitted line should pass through the point (x_0, y_0). We have therefore to impose this condition on the line and choose values for the parameters in such a way as to minimise the sum of the squared distances from the n given points to the fitted line where these distances are measured perpendicular to the x-axis or perpendicular to the fitted line as appropriate.

This fitting problem is illustrated in Figure 4.10 where we have employed the same data as in Figure 4.8. The only difference is that the fitted line is now constrained to pass through the point $(x, y) = (2, 3)$ corresponding to the constraint $a + 2b = 3$.

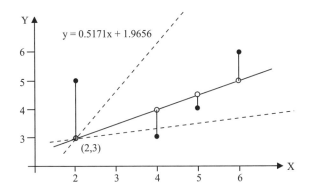

Figure 4.10 Fitted line constrained to pass through a given point.

4.6.2   Hypothesis Test

Substituting the condition $a = y_0 - bx_0$ in the optimality function, we obtain the constrained potential energy function

$$P_C = \frac{1}{2}\sum [y_i - y_0 - b(x_i - x_0)]^2$$

This expression is minimised when b is set equal to

$$b^\# = \frac{\sum (x_i - x_0)(y_i - y_0)}{\sum (x_i - x_0)^2}$$

and has a minimum value of

$$P^\# = \frac{1}{2}\sum (y_i - y_0)^2 - \frac{1}{2}\frac{[\sum (x_i - x_0)(y_i - y_0)]^2}{\sum (x_i - x_0)^2}$$

Similarly, for the minimum value of the unconstrained optimality function, we have to set $b = \hat{b}$ and $a = \bar{y} - \bar{x}\hat{b}$ to obtain the minimum value of the unrestricted potential energy function

$$P^* = \frac{1}{2}\sum (y_i - \bar{y})^2 - \frac{1}{2}[\sum (x_i - \bar{x})(y_i - \bar{y})]^2 / \sum (x_i - \bar{x})^2$$

The value of the potential energy function associated with the optimal solution to the constrained problem will necessarily be no smaller than that associated with the optimal solution to the unconstrained problem. Thus, the difference in the potential energy levels $P^\# - P^*$ may be used as the basis of a natural test of the hypothesis that the slope and intercept parameters truly satisfy the constraint $a + bx_0 = y_0$.

Now, the value of the function $P^\# - P^*$ depends on the scale of y (or, rather, on the unknown scale of the error term embodied in y) and we have to remove this unknown scale parameter before we can employ this difference in potential energy levels as the basis of a statistical test. One method of achieving this end is to divide the difference by the average unconstrained value P^*/n to yield the Wald statistic

$$n(P^\# - P^*)/P^*$$

A second approach is to divide this function by the average con-strained value $P^{\#}/n$ to yield the score (or Lagrange Multiplier) statistic

$$n(P^{\#} - P^{*})/P^{\#}$$

where this statistic is named for the partial derivative or force functions of Section 4.3. And, a third alternative is to employ a statistic (known as the likelihood ratio statistic) based on the dif-ference between the logarithms of the potential energy functions

$$n\log(P^{\#}/P^{*}) = n\log(P^{\#}) - n\log(P^{*})$$

Whichever of these three principles is adopted, the associated statistical test is performed by comparing the numerical value of the chosen statistic with a suitable critical value, usually the upper 1 or 5 percent critical value of the chi-squared distribution with one degree of freedom. If the numerical value of the statistic is larger than the chosen critical value, then the hypothesis under the test should be rejected.

The effect of imposing the constraint $a + 2b = 3$ on the para-meters of the line fitted to the data given in Table 2.1 is illustrated in Figure 4.11. With the imposition of this constraint, the fitted line of Figure 4.8 (representing the unconstrained minimum of the optimality function) has to be moved against the force in the stretched springs until it takes up a new position passing through the point $(x, y) = (2, 3)$. The stronger the resistance of the springs,

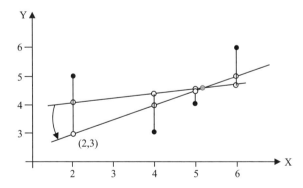

Figure 4.11 Fitted line forced to pass through a given point.

the less likely is the data to satisfy the constraint and the larger will be the value of the associated test statistic.

These diagrams also give some indication of the degrees of freedom of the problem, as the unconstrained line of Figure 4.8 is free to move in the two-dimensional xy-plane whilst the constrained line of Figure 4.10 is obliged to rotate about a fixed point in this plane. Thus, the unconstrained line has two degrees of freedom whereas the constrained line has a single degree of freedom. Further, the imposition of the constraint is associated with the loss of one degree of freedom and it seems appropriate that, under suitable conditions, all three tests of the validity of this hypothesis should have a chi-squared distribution with one degree of freedom.

4.6.3 Slope Test

The simple mechanical model described in Subsection 4.6.2 runs into a certain amount of difficulty if the linear constraint of interest does not involve the intercept parameter, but simply specifies a value $b = b_0$ for the slope parameter. In this context, we may impose this constraint approximately by choosing an arbitrarily large value for x_0 and setting $y_0 = b_0 x_0$. Although, in principle, this approach may be adopted when actually conducting a test, it seems more appropriate to impose the constraint explicitly to obtain the function

$$P_C = \frac{1}{2}\sum [y_i - a - x_i b_0]^2$$

which is minimised when a is set equal to

$$a = \bar{y} - \bar{x}b_0$$

so that the minimum value of the constrained optimality function for use in the tests of the hypothesis $H_0 : b - b_0$ is given by

$$P^{\#} = \frac{1}{2}\sum [y_i - \bar{y} - b_0(x_i - \bar{x})]^2$$

In either case, the value of the test statistic is again associated with the quantity of potential energy needed to force the fitted line to adopt a slope of $b = b_0$.

4.7 Inertial Models

In this chapter, we have outlined a convenient model for the method of least squared deviations based on a set of springs. An alternative mechanical model for the fitting of a line by the method of orthogonal least squares was discussed by Karl Pearson in 1901. The points in the plane representing observations on the variables X and Y are replaced by point masses of unit weight which are rigidly embedded at the relevant positions on a weightless physical plane. Arbitrarily selecting a line in the plane as axis of rotation, this rigid structure is rotated about the axis and the corresponding moment of inertia (or second moment) determined. In the context of this model, the problem is to determine the axis of rotation of the weighted plane in such a way as to minimise the moment of inertia of this system. The optimal position of this axis determines the orthogonal least squares fit of the line.

An important special case of this inertial model occurs when we wish to fit a single point to a set of observations on a single variable. Again, representing the n observations as point masses rigidly embedded in a straight line, we may rotate the weighted line about an arbitrary point $y = a$ on the line to obtain the moment of inertia for the system

$$2P = \sum (y_i - a)^2$$

This function is again minimised when its scaled first derivative

$$\frac{dP}{da} = - \sum (y_i - a)$$

is set equal to zero.

This optimality condition again defines the arithmetic mean of the observations, and the moment of inertia model described in this section may be associated with the balance of rotational forces model developed in Chapter 1. However, it must be stressed that there is no clear mechanical relationship between the two. In particular, we note that all n observations in the inertial model travel in the same (clockwise) direction about the arbitrary point whereas, in the case of the chemical balance model of Chapter 1, some of the points will tend to move in a clockwise direction whilst others will tend to move in the opposite direction.

Although this inertial model clearly has its merits, it lacks the simplicity of the potential energy models based on springs, and we shall not discuss it further in this book.

4.8   Matrix Representation

The least squares problems discussed in this chapter, and the corresponding traditional procedures of multivariate analysis discussed in Chapter 11, may readily be expressed in matrix notation. Although we have made some use of this notation in the present chapter as it serves to express algebraic results easily and to point up references to the earlier knowledge of some readers, we have not made excessive use of the facility as to do so would draw an unnecessary distinction between fitting procedures based on the method of least squares and those based on other optimality criteria.

Although there may have been some justification for practitioners restricting themselves to statistical procedures with simple algebraic forms in the era before the advent of modern high-speed computers, this restriction ceased to be justified once such equipment and the associated computing packages had become readily available.

References

Courant, R. (1940), Soap film experiments with minimal surfaces, *American Mathematical Monthly* **47**: 167–174.

Courant, R. and H. Robbins (1941), *What is Mathematics?*, Oxford University Press, New York.

Donkin, W. F. (1844), An essay on the theory of combination of observations, *Transactions of the Ashmolean Society* **2** (18): 1–71.

Farebrother, R. W. (1999), *Fitting Linear Relationships: A History of the Calculus of Observations* 1750–1900, Springer-Verlag, New York.

Hald, A. (1998), *A History of Mathematical Statistics from 1750 to 1930*, John Wiley and Sons, New York.

Pearson, K. (1901), On lines and planes of closest fit, *Philosophical Magazine*, Series 6, **2**: 559–572.

Sall. J. (1991), The conceptual model behind the picture, *ASA Statistical Computing and Statistical Graphics Newsletter* **2**: 5–8.

Stigler, S. M. (1986), *The History of Statistics: The Measurement of Uncertainty Before 1900*, Harvard University Press, Cambridge, Massachusetts.

CHAPTER 5

Method of Least Absolute Deviation

5.1 One- and Two-Dimensional Medians

5.1.1 Mechanical Models for One-Dimensional Medians

In this chapter, we shall resume the analysis of the least sum of absolute deviations problem introduced in Chapter 3. As a reminder of some of the material discussed in that chapter, we again specify the potential energy function associated with the position of a single point. Suppose that we are given a horizontal board. Then we may draw an arbitrary straight line on the board and identify this line with the y-axis of a Cartesian system. Arbitrarily selecting a point on this line, we may drill a hole through it and pass a length of string through this hole. We attach a unit weight to the lower end of the string which thus hangs vertically from the hole whilst the portion of the string that is above the board is pulled taut along the y-axis. As the upper end of the string is pulled along the y-axis the weight either rises and thus gains potential energy at a steady rate, or falls and loses potential energy at the same steady rate.

Denoting the position of the upper end of the string by $y = a$, and plotting the potential energy as a function of this parameter, we find that the potential energy function associated with this simple model takes the form of two lines drawn at 45 degrees to the horizontal a-axis $P_i = |y_i - a|$, see Figure 5.1b.

The derivative of this function clearly takes the value -1 when a is less than y_i, the value $+1$ when a is greater than y_i, and is undefined when a is equal to y_i. There is thus some difficulty in extending the simple differential analysis of the least squares problem to this case.

Now, suppose that instead of a single point on the y-axis, we have a scatter of points corresponding to a set of n observations on a single variable. Constructing a horizontal line as before, we may drill a hole through the board at each of the locations corresponding to observations and pass a length of string through each of these holes. A unit weight is attached at the lower end of each of these strings. The upper ends of the strings are jointly tied to a single ring at an arbitrary point on the y-axis. As the ring is moved along this axis the weights rise and fall as before, thus defining the potential energy functions of the individual weights. The sum of these individual functions defines the piecewise linear potential energy function of the system as a whole, see Figure 5.2b. The minimum value of this overall potential energy function identifies the median of the data set as we shall now show.

Returning to the sum of the individual functions, we may re-examine the problem on the basis of the balance of forces in the system. For ease of exposition, we shall, as usual, suppose that the y-axis runs from south to north through the horizontal plane. If

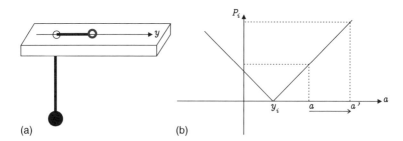

(a)   (b)

Figure 5.1 (a) The physical model, and (b) the potential energy function associated with a single observation.

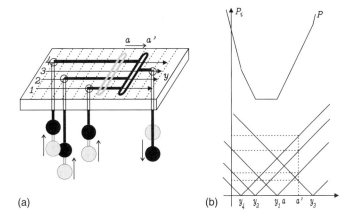

Figure 5.2   (a) Balance of forces, and (b) the potential energy functions associated with a set of observations.

there are r observations to the south of the current value and $n - r$ to the north of this value, then there will be $n - r$ strings pulling the ring northwards and r strings pulling it southwards. Thus there is a net force of $n - 2r$ units pulling the ring northwards, and the ring will move in this direction if n is greater than $2r$ and southwards if n is less than $2r$.

   This argument is illustrated in Figure 5.3. As in Section 3.1, we have again separated the forces pulling the ring in the south–north direction by associating arbitrary x values with the given y values and by stretching the circular ring in the west–east direction to obtain an oval with straight sides.

   If n is an odd number, then the ring will be at rest if it is located at the mth observation where $m = (n + 1)/2$. Similarly, if n is even and $m = n/2$ then the ring will be at rest at any point on the interval between the mth and the $(m + 1)$th observations. As is well known, these two properties characterise the median of a set of univariate observations.

   Setting s_i equal to the force in a northwards direction in the ith string, we find that we may identify this function with the negative of the derivative of the ith contribution to the potential energy function $|y_i - a|$ except when a is equal to y_i, when the

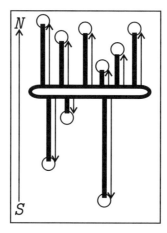

Figure 5.3   Balance of forces in the south–north direction.

derivative is undefined. This exception is reasonable as s_i can take any value between -1 and $+1$ (inclusive) in this case. A consideration of the definition of the median when $n = 2m - 1$ is odd, and there are two or more values of y_i taking the same value as y_m, will help to demonstrate this point. For example, if there are $r + 1 \geq 2$ such values with indices $i = m$, $m + 1,\ldots,$ $m + r$ or $i = m - r$, $m - r + 1,\ldots,$ m, then we have to share r units of force between these $r + 1$ strings, and we do so by setting $s_i = r/(r + 1)$ for $i = m$, $m + 1,\ldots,$ $m + r$ or $s_i = -r/(r + 1)$ for $I = m - r$, $m - r + 1,\ldots,$ m respectively.

5.1.2   Mechanical Models for Higher-Dimensional Medians

In Chapter 3, we generalised our one-dimensional model to the two-dimensional case by simply removing the restriction that the holes drilled through the board must lie on a straight line. In principle, this model may be further generalised to higher dimensions by replacing the holes drilled through a physical board by a set of eyelets suspended in space. In this context, suppose that we are given a set of n observations on a set of $q \geq 2$ variables. Then these n observations may be represented by n points in q-dimensional space. Our problem becomes one of choosing an additional point in q-dimensional space in such a way that the sum of the absolute distances from this additional point to the n given points is minimised.

Generalising the mechanical model of Subsection 5.1.1 to this case, we find that we have to place an eyelet at each of the n points indicated by the data, pass n strings with unit weights at their free ends through these n eyelets, before tying their upper ends to a ring located at the additional point. Assuming that the weights are attracted by a force acting at right angles to the q-dimensional space of observations, we find that the potential energy of the system as a whole is again proportional to the sum of the absolute distances, and thus that the mediancentre is defined by the minimum value of this function.

5.2 Simple Linear Regression

5.2.1 Conventional Line Fitting Problem

As in Chapter 4, we have to generalise the mechanical model for fitting a single point to a set of observations to the problem of

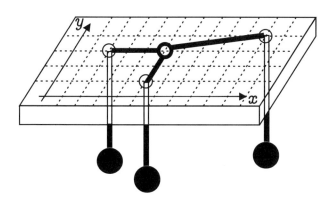

Figure 5.4 Mechanical model for the two parameter case.

fitting a straight line in such a way as to minimise the sum of the absolute distances from this line to the given points. As before, we have to associate a real horizontal plane with the abstract Cartesian plane of observations, then at each point (x_i, y_i) associated with an observation, we have to drill a hole. Through this hole we pass a length of string with a unit weight at the lower end and a small ring at the upper end. We now place a rigid rod at an arbitrary position in the plane and pass the n small rings over this rod in such a way that the associated strings run in a south–north direction from the ring to the hole. Now, each of the weights induces a unit force on the rod tending to pull it northwards or southwards towards the hole. Separating the forces tending to pull the rod northwards from those tending to pull it southwards, we may determine whether the net force acting on the rod will tend to pull it northwards, to pull it southwards, or to leave it in its present position.

Besides the north–south forces acting on the rod, we also have to consider its tendency to rotate about a particular point on the rod. Arbitrarily selecting such a point, we have to evaluate the distance measured parallel to the x-axis in a west-east direction from each of the observed data points to a south–north line through this arbitrary point. If the ith data point lies to the west of the arbitrary point and to the north of the rod, or to the east of the arbitrary point and to the south of the rod, then the unit force in the ith string will pull the rod in a clockwise direction about the selected arbitrary point. Multiplying the unit force in the ith string

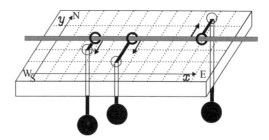

Figure 5.5   Observation space representation of the conventional line fitting problem.

by the perpendicular distance between this string and the arbitrary point, we obtain the associated rotational force or couple acting in a clockwise direction. Summing these clockwise couples and comparing the result with the corresponding total for the anticlockwise direction, we may determine whether the forces in the weighted strings will cause the rod to rotate about the chosen arbitrary point in a clockwise direction or in an anticlockwise direction.

Clearly, this system is in equilibrium when the forces tending to pull the rod northwards are exactly balanced by those tending to pull it southwards, and the couples tending to turn the rod in a clockwise direction about the arbitrary point are exactly balanced by those tending to turn it in an anticlockwise direction.

Defining the deviation of the ith point from the line $y = a + bx$ by $e_i = y_i - a - bx_i$, and taking couples about the point $x_0, a + bx_0)$ on the line, we may express this result in the following algebraic terms: $\sum_i |e_i|$ is minimised when $\sum s_i = 0$ and $\sum(x_i - x_0)s_i = 0$, that is when $\sum s_i = 0$ and $\sum x_i s_i = 0$, where s_i represents the force in the ith string and takes the value $+1$ when e_i is positive, the value -1 when it is negative, and a value lying between these limits (inclusive) when e_i is zero. It is tempting to refer to s_i as the sign of e_i, but this usage is potentially misleading and should be resisted. [Note that the value of x_0 used to define the position of an arbitrary point on the fitted line does not feature in the reduced set of equations characterising the solution to this problem.]

The equilibrium of forces in the south–north direction when there are an odd number of observations means that the line must

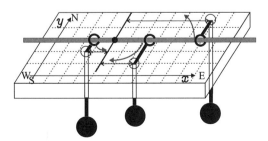

Figure 5.6 Balance of rotational forces in the line fitting problem.

pass through at least one of the data points. In a similar way, the corresponding equilibrium of clockwise and anticlockwise couples about the selected data point implies that the line usually passes through two data points. Of course, there are exceptional circumstances, but this characterisation of a potential solution to the least sum of absolute deviations problem holds true in general.

For example, suppose that we are given the $n = 3$ points $(x_1, y_1) = (0, 4)$, $(x_2, y_2) = (4, 0)$ and $(x_3, y_3) = (5, 8)$. Then the fitted line will either pass through the first and second of these points, thereby defining the line $y = 4 - x$ with the third point at a distance 9 units from the line; or it will pass through the first and third of these points, thereby defining the line $y = 4 + 0.8x$ with the second point at a distance 7.2 units from the line; or it will pass through the second and third of these points, thereby defining the line $y = 8x - 32$ with the first point at a distance 32 units from the line. The second of these options clearly gives the smaller value for the sum of absolute deviations optimality function, and we may deduce that the optimal line passes through the first and third of the given points.

Indeed, we could have reached the same conclusion without the need for any calculations as the first option is associated with a net anticlockwise couple about the second point, so that the fitted line is not in equilibrium, but will tend to rotate in an anticlockwise direction about this point; similarly, the third option is associated with a net clockwise couple about the second point, so that the fitted line will tend to rotate in a clockwise direction about this point.

In the particular case in which there are three noncollinear points in the plane and the rod passes through the first and third of these points, there will be a unit force acting on the rod tending to pull it towards the second point. Further, since the rod is in a state of equilibrium, the unit force tending to pull the rod towards the second point must be balanced by a unit force tending to pull it in the opposite direction. This balancing force must be supplied by the weights hanging vertically from the rod through the pair of holes at the points defining the current position of the rod. That is, each of the strings hanging vertically from the rod must be associated with a horizontal force tending to pull the rod in a direction that is opposite to the force pulling it towards the second point. Further, the couples associated with these forces taken about the point at which the offline string is attached to the rod must balance. Thus, the point at which this string is attached to the rod must lie between the two points determining the position of the rod, and the horizontal forces associated with these two points must be chosen in such a way that the couples about this point balance, see Figure 5.7.

Now, suppose that the three points in this problem have coordinates (x_1, y_1), (x_2, y_2) and (x_3, y_3) with $x_1 < x_2 < x_3$, then the position of the fitted line will be determined by the first and third of these points, and there will be a unit force pulling the fitted line towards the point (x_2, y_2). Balancing this unit force, and acting in the opposite direction, there will be a force of $(x_3 - x_2)/(x_3 - x_1)$ at the point (x_1, y_1) and a force of $(x_2 - x_1)/(x_3 - x_1)$ at the point (x_3, y_3). Where the numerators of these expressions have been chosen in such a way that the couple acting at a distance (measured parallel to the x-axis) $x_2 - x_1$ from the south–north line through the point (x_2, y_2) exactly balances the second acting at a distance $x_3 - x_2$ from this line.

For example, if we again consider the case in which we are given the points $(x_1, y_1) = (0, 4)$, $(x_2, y_2) = (4, 0)$ and $(x_3, y_3) = (5, 8)$, then the optimal line will pass through the first and third of these points, thereby defining the line $y = 4 + 0.8x$. The second point is attached to this line at $(x, y) = (4, 7.2)$ and will

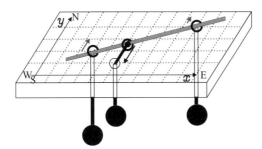

Figure 5.7   Position of the rod determined by three points.

be associated with a unit force pulling the line southwards, whereas the point at $(0, 4)$ will be associated with a northwards force of 0.2 and the point at 5, 8 with a northwards force of 0.8.

At first sight, it might be thought that the conceptual difficulty of associating horizontal forces with weights hanging vertically from a rod may be overcome by moving the rod a small distance parallel to itself towards the third point. However, this simple solution is a mirage as moving the rod in this way will establish a pair of unit forces tending to pull the rod towards its original position.

We do not propose to discuss the weighted sum of absolute deviations criterion in this chapter. However, it should be noted that our discussion of the horizontal effects of weights hanging vertically from a rod may have nontrivial implications for Koenker and Basset's (1978) concept of regression quantiles. For each value of k in the range $0 < k < 1$, the kth regression quantile is defined by the values of a and b which minimise the weighted sum of the absolute deviations

$$\sum w_i |y_i - a - bx_i|$$

where w_i takes the value k if the ith observation lies on the fitted line, or is to the north of it, $(e_i \geq 0)$, and it takes the value $1 - k$ if it is (strictly) to the south of the fitted line $(e_i < 0)$.

5.2.2 Orthogonal Line Fitting Problem

Having discussed the line fitting problem in the conventional case, the corresponding exposition for the line which minimises the sum of the orthogonal deviations is readily obtained by treating the fitted line as though it were fixed in a west–east direction. The plot of points in the xy-plane is then adjusted to this fixed line in such a way that the sum of the distances to the given points is minimised where distances are measured perpendicular to the fitted line, that is, in the south–north direction. Although this geometrical description of the problem is relatively simple, it will be apparent from our discussion of the orthogonal least squares problem in

Section 4.3 that the algebraic solution of the orthogonal least sum of absolute deviations problem is far less so, and we shall therefore not discuss it further in this book.

5.2.3   Linear Constraints

As a variant of the unconstrained problem, we may choose values for the parameters a and b in such a way as to minimise the sum of the absolute deviations $\sum |e_i|$ subject to a linear constraint which defines a in terms of b. This linear constraint requires that the line passes through a given point in the xy-plane. For example, if we know that the parameters satisfy the constraint $a = y_0 - bx_0$, then the fitted line must pass through the point (x_0, y_0). Since the line is not able to move parallel to itself, we have only to establish an equilibrium between the clockwise and anticlockwise couples about this fixed point. This approach to the line fitting problem was taken by Rogerius Josephus Boscovich (Croatian: Rudjer Josip Bošković) in his original formulation of the constrained least sum of absolute deviations problem in 1760, some forty-five years before Adrien-Marie Legendre announced the solution of the unconstrained least squares problem. In his original formulation of this problem, Boscovich chose values for a and b in such a way as to minimise the sum of the absolute deviations subject to the condition that the corresponding signed deviations should sum to zero. Now, the condition $\sum e_i = 0$ implies that $\sum y_i = na + b \sum x_i$, and hence that the fitted line must pass through the centroid $(\overline{x}, \overline{y})$ of the observations on X and Y. Thus, Boscovich had only to choose a value for b in such a way that the clockwise couples about the centroid balanced the anticlockwise couples about this point.

   Plotting the observations on a Cartesian plane and drawing a movable line through the centroid, he set b at an arbitrarily large positive value so that the initial position of the candidate line represented a small clockwise rotation of the south–north line through the centroid. He then gradually rotated this line in a clockwise direction and evaluated the optimality function at each encounter with a data point until the sum began to increase

again. The point associated with the minimal value of the optimality function was then identified as the optimal value of b and the
corresponding value of $a = \bar{y} - b\bar{x}$ determined from the adding-up
constraint $\sum e_i = 0$.

This procedure is illustrated in Figure 5.8. The points
$(x, y) = (5, 8)$, $(4, 0)$, and $(0, 4)$ are plotted in the Cartesian plane
together with their centroid, $(x, y)_= = (3, 4)$. As the moving line
rotates in a clockwise direction about the centroid, it passes
through these three points in succession: through the point $(x, y) =$
$(5, 8)$ when its slope is $b = 2$, through the point $(x, y) = (0, 4)$ when
its slope is $b = 0$. And through the point $(x, y) = (4, 0)$ when its
slope is $b = -4$. Now, the parameters satisfy the relationship
$a = 4 - 3b$, so that the constrained optimality function may be
written as

$$\sum |e_i| = |4 - 2b| + |3b| + |4 + b|$$

and this function takes the values 12, 8, and 24 when b takes the
values 2, 0, and -4. We may therefore deduce that the optimal
constrained values of the parameters are $b = 0$ and $a = 4$ corresponding to the line $y = 4$.

This geometrical form of the constrained least sum of absolute
deviations fitting procedure was subsequently popularised in an
algebraic form by Laplace, see Farebrother (1999, pp. 25–29) or
Stigler (1986, pp. 50–55) for details. The unconstrained form of
this problem and the realisation that its solution corresponds to a
hyperplane passing through as many points as there are parameters in the problem were due to Carl Friedrich Gauss.

In passing, we note that Boscovich's choice of the adding-up
constraint $\sum e_i = 0$ has nothing to do with the least squares criterion (which was first proposed several years after Boscovich's
death) although of course they both derive from the same statistical considerations.

5.2.4 Hypothesis Tests

In earlier sections, we have seen that the sum of the absolute
deviations function may be interpreted as a potential energy func-

tion and that the constrained optimal point will be associated with a larger value of this function than is the unconstrained optimal point. It therefore seems reasonable to suggest that we may use this difference in potential energy levels as the basis of a natural test of the hypothesis that the parameters of the underlying model truly satisfy the selected constraint. However, unfortunately, this was not the approach adopted by Koenker and Bassett (1982) and Koenker (1987) in their analysis of the subject. Instead, they established the asymptotic distributions of three variants of the statistics defined in Section 4.6 and we refer interested readers to these articles for further details.

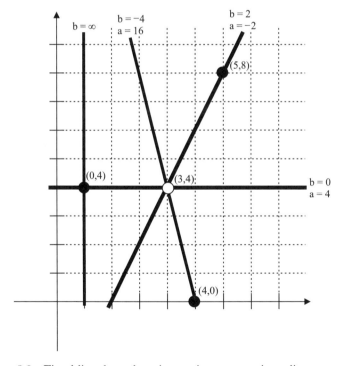

Figure 5.8   Fitted line through a given point representing a linear constraint.

5.3   Parameter Space Representation and Duality

5.3.1   Simple Mechanical Model in Parameter Space

As briefly explained in Section 2.3, instead of representing the ith pair of observation $X - x_i$ and $Y = y_i$ as the point (x_i, y_i) in the xy-plane of observations, we may prefer to represent it as the line $a = y_i - bx_i$ in the ba-plane of slope and intercept parameters. Similarly, instead of representing the fitted relationship as the line $y = a + bx$ in the xy-plane, we may prefer to represent it as the point (b, a) in the ba-plane. Thus, points in the xy-plane are directly associated with lines in the ba-plane and *vice versa*, see Figure 5.9. Further, a line in either system is defined by the line segment joining a pair of points, and a point by the intersection of a pair of lines. This dual relationship between the two systems of representation in the xy-plane of observations and the ba-plane of parameters is entirely natural (if seldom mentioned) in statistics, but it appears as a distinct subject area in mathematics. This formal subject area, known as Projective Geometry, developed from the realisation in the early Renaissance that there was a simple mathematical theory underlying the graphical representation of perspective in art.

Plotting the ith observation as the line $a = y_i - bx_i$ in the ba-plane, we note that the arbitrary point with coordinates (b, a) is at a distance $e_i = y_i - a - bx_i$ (measured parallel to the a-axis) from the point $(b, y_i - bx_i)$ on this line. Thus, we may readily determine the amount by which this observation deviates from a fitted relationship associated with a particular choice of parameter values by measuring the distance parallel to the a-axis between the relevant point (b, a) and the ith line in the plot of lines.

Thus, we may readily obtain a simple mechanical model for the line fitting problem in the space of parameters by passing a weighted string over each of the lines and attaching them to a ring at an arbitrary point in the ba-plane, see Figure 5.10. If all of these strings carry unit weights and are constrained to lie parallel to the

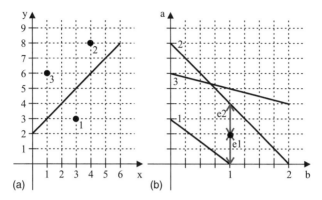

Figure 5.9 Representation of three observations and a linear relationship (a) in observation space, and (b) in parameter space.

a-axis, then the potential energy of the system of weights is equal to the sum of the absolute distances $\sum |e_i|$ and we only have to find a position for the ring to minimise this optimality function.

5.3.2 Perpendicular Models in Parameter Space

The parameter space representation of the line fitting problem developed in subsection 5.3.1 assumes that the strings are constrained to lie parallel to the a-axis. A more attractive model may be obtained by permitting the strings to lie perpendicular to the individual lines constructed in the ba-plane.

If the arbitrary point is located at the point (b, a) in the ba-plane and the ith line is defined by the equation $a = y_i - bx_i$, then we have seen that the arbitrary point is at a distance $e_i = y_i - a - bx_i$ from this line when distances are measured parallel to the a-axis. Further, since the given line has a slope of $-x_i$, the perpendicular to this line must have a slope of $1/x_i$, and reference to Figure 5.11 shows that the distance $e_i = y_i - a - bx_i$ measured parallel to the a-axis may be resolved into a distance of $e_i/\sqrt{(1 + x_i^2)}$ measured perpendicular to the ith line and a distance of $x_i e_i/\sqrt{(1 + x_i^2)}$ measured parallel to this line.

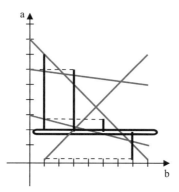

Figure 5.10   Simple mechanical model in parameter space.

The alternative mechanical model for the least absolute deviations fitting problem is obtained by permitting the n strings to lie perpendicular to the n given lines. In this context, we have to attach a weight proportional to $\sqrt{(1 + x_i^2)}$ to the lower end of the length of string passing over the rod representing the ith line, see Figure 5.12a.

The potential energy of this system of weights is proportional to the weighted sum of the absolute perpendicular distances from the ring to these n lines, and this choice of weights ensures that the potential energy function is proportional to the

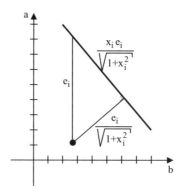

Figure 5.11   Resolution of a vector into orthogonal components.

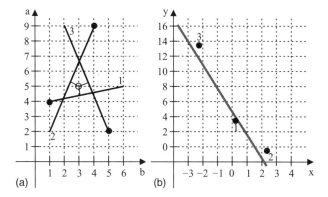

Figure 5.12   (a) Mechanical model with distances measured perpendicular to the observed lines in parameter space, and (b) the corresponding fitted line in observation space.

unweighted sum of the absolute distances (measured parallel to the a-axis) from the ring to the given lines. In equilibrium, this mechanical system will therefore determine the values of a and b which minimise the unweighted sum of the absolute distances between the arbitrary point in parameter space and the given set of lines, or equivalently, the sum of the corresponding distances in observation space between the arbitrary line to the given set of observations, see Figure 5.12b. Both of these mechanical models in parameter space may readily be generalised to the least squares problem of Chapter 4 by replacing the ith string by a spring of modulus unity in the first case and modulus $1 + x_i^2$ in the second. Indeed, the (primal) mechanical model of Section 5.2 and the second of these (dual) mechanical models were developed in the context of the least squares problem; the latter being proposed by Donkin in 1844 and the former by Newcomb in 1873, see Farebrother (1999, pp. 168–171) for details. Further, it should be noted that Donkin's mechanical model was developed in the context of a three-dimensional problem, so that observations were represented by two-dimensional planes and the potential energy functions associated with the fitted point were proportional to the squared perpendicular distances from this point to the given planes.

5.3.3 Mechanical Models for Directional Data

The second mechanical model of Subsection 5.3.2 arises naturally when the data consist of a set of observations on the angular directions or bearings of a single object measured from n known locations. If the ith observation records that the object lies in a direction which is θ_i degrees to the east of north when viewed from a point with known map coordinates (x_i, y_i), then we have to draw a line in the xy-plane which passes through the point (x_i, y_i) with slope $b_i = \tan(\theta_i)$. That is, we have to draw the lines $y = y_i + (x - x_i)b_i$ for $i = 1, 2, \ldots, n$. In practice, these n lines will not intersect in a single point and our problem becomes one of determining a point (x_0, y_0) that minimises some increasing function of the perpendicular distances from this point to the n given lines.

Translating this problem to the ba-plane, we find that the n lines of observations are represented by points $(b_i, y_i - x_i b_i)$ for $i = 1, 2, \ldots, n$, and the arbitrary point by a line $a = y_0 - bx_0$. We shall not discuss this problem at greater length, but it is clear that mechanical models for this problem may readily be developed from those discussed earlier in this chapter.

5.4 Geometrical Solution

5.4.1 Geometrical Plot of the Optimality Function

Returning to the first model of Section 5.3, we have to explain how the three-dimensional figure representing the sum of absolute deviations optimality function may be constructed from the plot of lines in the ba-plane of Figure 5.9b. Our earlier argument established that the ith observation makes a contribution to the overall potential energy function which is given by the distance of the selected point from the ith line $a = y_i - bx_i$ in the ba-plane. Thus, for each line in this plane, we have to construct a pair of half planes terminating in this line and rising from the horizontal plane at an angle of 45 degrees (where this angle is measured parallel to the a-axis). The uppermost surface of the figure determined by the complete set of half planes takes the form of a mould for a polyhedral vase and defines the optimality function for the minimax absolute deviation problem of Chapter 6. In the context of the least sum of absolute deviations problem, the relevant optimality function is obtained by summing the values given by the individual potential energy functions. The resulting optimality function again takes the form of a mould for a polyhedral vase, but with many more facets than in the case of the minimax problem, so that the product of this mould resembles a sugar bowl rather than a whisky tumbler. The optimality function thus consists of a set of plane sections joined together along edges which lie vertically above the lines plotted in the ba-plane of Figure 5.9b. In either case, the lowest point of the relevant piecewise planar function determines the parameters of the optimal choice of a fitted line.

Unfortunately, it has not proved possible to produce a convincing two-dimensional representation of the three-dimensional sum of absolute deviations optimality function. The best we are able to offer is a shading of the different areas of Figure 5.9b in such a way as to indicate the relative slopes of the corresponding facets of the polyhedral vase. It is interesting to note that this figure was mentioned implicitly by Fourier in an aside to his discussion of the minimax absolute deviations problem in 1827. It was first explicitly described by Francis Ysidro Edgeworth in 1888 when he noted that its negation resembled "the roof of an irregularly built slated house".

5.4.2 Edgeworth's Double Median Method

A first method for determining the optimal values of the parameters in parameter space is suggested by the first mechanical model of Subsection 5.3.2. In this context, we have to plot the n lines $a = y_i - bx_i$ for $i = 1, 2, \ldots, n$ in the ba-plane. Then we have to choose a fixed value for b, draw the corresponding line parallel to the a-axis in the plot of lines, and read off the values of $y_1 - bx_1, y_2 - bx_2, \ldots, y_n - bx_n$. The optimal value of a conditional on this choice of b is clearly the middlemost of these n values. Thus, the optimal value of a for each choice of b is easily found when n is odd, by identifying the middlemost line in the plot of lines corresponding to this particular choice of a value for b. On the other hand, if n is even, then an optimal value for a is given by any point on the line segment lying between the two middlemost lines.

To find the set of all such conditional solutions in a systematic manner when n is odd, we begin by identifying the middlemost line corresponding to an arbitrarily large and negative value of b. As b gradually increases, the optimal solution will move along this middlemost line until it intersects with another line and the optimal solution transfers to a new middlemost line. In this way, we obtain the trace of a broken line running from west to east through the diagram which records the optimal choice of a value for a conditional on a particular choice of a value for b, see Figure 5.13a. If, on the other hand, n is even, then this connected sequence of line segments will be replaced by a connected sequence of quadrilaterals as illustrated in Figure 5.13b.

For simplicity, we shall restrict our discussion of the problem to the case when n is odd. In this case, the solution to the least sum of absolute deviations problem may be determined by evaluating

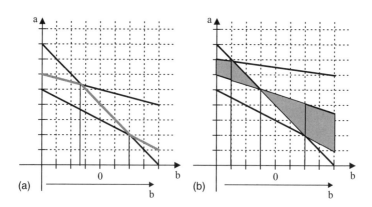

Figure 5.13 (a) Sequence of line segments, and (b) sequence of quadrilaterals identifying the optimal value of a for different choices of b.

the sum of the absolute deviations function $\sum |y_i - a - bx_i|$ for a range of values of a and b lying on the broken line of a-medians. The minimum value of this function naturally corresponds to the solution of the problem. Although this technique works well enough for simple problems such as the one illustrated here, it seems sensible to reduce the search area as much as possible before initiating the search itself.

An approximate constraint on the set of solutions may be obtained by plotting a similar broken line of b-medians (actually pseudo-b-medians) running from south to north through the diagram and considering those points lying on the intersection of the two lines of medians. Although this approximate procedure (first proposed by Arthur Lyon Bowley in 1906) may well give satisfactory results in practice, it is strictly incorrect as it ignores the weighting implicit in the second line of medians. Further, there is little justification for using this approximate technique nowadays as it is relatively easy to obtain a plot of the true b-medians.

Given a particular value of a, we have to choose b to minimise the sum of the absolute deviations function $\sum |y_i - a - bx_i|$. Setting the zero values of x_i on one side, we find that the adjusted function may be written as

$$\sum |y_i - a - bx_i| = \sum |x_i||(y_i - a)/x_i - b|$$

so that we have to set b equal to the weighted median of the ratios

$$(y_1 - a)/x_1, (y_2 - a)/x_2, \ldots, (y_n - a)/x_n$$

excluding those expressions for which $x_i = 0$.

For any given value of a, we renumber the equations in such a way that these ratios are in increasing order and evaluate the corresponding sequence of partial sums $|x_1|, |x_1| + |x_2|, \ldots, \sum |x_i|$, where the subscripts now refer to this new ordering of the observations. We then set b equal to the value $(y_j - a)/x_j$ where this ratio is chosen in such a way that the $(j - 1)$th partial sum $|x_1| + \ldots + |x_{j-1}|$ is strictly less than one-half of the total $\sum |x_i|$ whilst the jth partial sum $|x_1| + \ldots + |x_j|$ is greater than, or equal to, this critical value. As before, we may define a second broken line running from south to north through the diagram by first setting a equal to a large negative value and then considering the effect of employing successively larger values of a. Such a line is plotted in Figure 5.14.

In Figures 5.13a and 5.14 we plotted three lines corresponding to the equations $A = -1 - b$, $A = -2 - 0.5b$, and $A = -0.25b$. For values of b less than $b = -8$ (not shown in these figures) the middlemost line is defined by the second of these equations. For values of b between $b = -8$ and $b = -1.33$ the middlemost line is defined by the third equation. For values of b between $b = -1.33$ and $b = 2$ the middlemost line is defined by the first equation. And, finally, for values of b greater than $b = 2$ the middlemost line is again defined by the second equation.

The standard calculations for identifying the line of b-medians are outlined in the table attached to Figure 5.14. However, there is no need to follow this procedure in the present case, as the absolute value of one of the x_i exceeds the sum of the other absolute values. In this context, the line of b-medians necessarily follows the line corresponding to the dominant x_i value. That is, the line of b-medians necessarily follows the line defined by the first equation. Combining these two sets of results, we find that the two lines of medians intersect on the line segment joining the point $(b, a) = (-1.33, 0.33)$ to the point $(b, a) = (2, -3)$.

If the line of a-medians intersects the line of b-medians in a single point, then this point of intersection necessarily identifies the

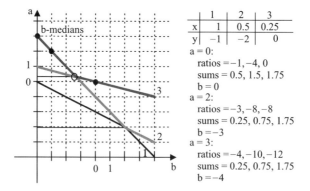

Figure 5.14 Sequence of line segments identifying the optimal value of b for given values of a.

unique solution to the fitting problem. However, in practice, these two lines of medians will often intersect in a sequence of one or more line segments and the first technique proposed by Edgeworth in 1888 will only identify a class of *possible* solutions to the fitting problem. Indeed, many of the points identified in this way will represent the optimal value of a conditional on b, and the optimal value of b conditional on a, without also being optimal solutions to the unconstrained problem. As before, we are obliged to investigate the value of the optimality criterion associated with a range of points on the common sequence of line segments if we wish to identify an optimal solution to the unconstrained problem. In Figure 5.15 and the associated table we have evaluated the sum of absolute deviations function $\sum |e_i|$ at nine points on the broken line of a-medians. Whether or not we restrict ourselves to the line segment common to the lines of a- and b-medians, we find that the smallest value achieved by this optimality function is 1.66, and hence we are able to deduce that the values $a = 0.33$ and $b = -1.33$ define the optimal least sum of absolute deviations line.

In principle, we may reduce the indeterminacy in this technique by determining the lines of weighted medians conditional on other linear combinations of a and b. But, in practice, the direct implementation of this approach is unlikely to produce a satisfac-

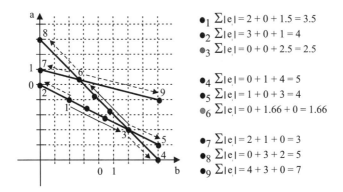

$\bullet_1$ $\sum |e_i| = 2 + 0 + 1.5 = 3.5$
$\bullet_2$ $\sum |e_i| = 3 + 0 + 1 = 4$
$\bullet_3$ $\sum |e_i| = 0 + 0 + 2.5 = 2.5$

$\bullet_4$ $\sum |e_i| = 0 + 1 + 4 = 5$
$\bullet_5$ $\sum |e_i| = 1 + 0 + 3 = 4$
$\bullet_6$ $\sum |e_i| = 0 + 1.66 + 0 = 1.66$

$\bullet_7$ $\sum |e_i| = 2 + 1 + 0 = 3$
$\bullet_8$ $\sum |e_i| = 0 + 3 + 2 = 5$
$\bullet_9$ $\sum |e_i| = 4 + 3 + 0 = 7$

Figure 5.15 Values of the optimality function at several points on the broken line of a-medians.

tory procedure. A more sophisticated implementation of this idea will be outlined in Subsection 5.4.4 below.

5.4.3 Edgeworth's Iterative Method

An alternative (iterative) approach to the sum of absolute deviations problem was also developed by Edgeworth in 1888. This second approach considers the value of the criterion function at selected points on the grid of lines in the ba-plane such as those illustrated in Figures 5.13a and 5.13b. Starting from any point on one of the oblique lines in this grid, he determined whether he should move in one direction or the other along the chosen line by considering the effect on the criterion function of a small change in the value of b. He then moved in the chosen direction to the first intersection between this line and another. Next, he considered the effect of moving in any one of the four directions along the pair of lines intersecting at this point, and so on until no further movement from a point of intersection will lead to a lower value of the criterion function.

Edgeworth did not explain why he restricted his attention to points of intersection in the plot of lines, but this is apparent from his description of the geometrical form of the criterion function given in Subsection 5.4.1.

5.4.4 An Improved Iterative Method

A much faster convergence to the optimal solution of the problem may be obtained by using the principle embodied in Edgeworth's double median method to make a minor modification to the iterative procedure of Subsection 5.4.3. Instead of moving from one point of intersection to the next in either direction along either of the pair of lines passing through the current point of interest, this modification considers the effect of moving from the current position to any other intersection on the lines passing through this point. Suppose that we have arrived along the hth line at its point of intersection with the jth line. Then this point of intersection identifies the lowest value of the optimality criterion available on the hth line, and we only have to consider whether a move to any other point on the jth line will lead to a reduction in the value of the optimality criterion. Now, we know that a is defined in terms of b on this line by $a = y_j - bx_j$, and we have to choose a value for b in such a way as to minimise the sum of the absolute deviations function

$$\sum_i |y_i - a - bx_i|$$

subject to this constraint. That is, we have to choose b to minimise the function

$$\sum_i |y_i - y_j - b(x_i - x_j)|$$

As in Subsection 5.4.2, we rewrite this expression in the form

$$\sum_i |[(y_i - y_j)/(x_i - x_j)] - b| |x_i - x_j|$$

and set b equal to the weighted median of the ratios

$$\frac{y_1 - y_j}{x_1 - x_j}, \frac{y_2 - y_j}{x_2 - x_j}, \ldots, \frac{y_n - y_j}{x_n - x_j}$$

omitting those expressions for which $x_i = x_j$.

This procedure sets b equal to one of the ratios defined by the intersection of the jth line with one of the other lines, say the kth line. We then take this as our new base line and look for a point of intersection which minimises the sum of the absolute deviations function subject to the constraint $a = y_k - bx_k$, and so on. This iterative procedure continues until no further reduction in the value of the optimality criterion is possible. This procedure has been programmed in Pascal by Farebrother (1988, 1992b).

5.4.5 Linear Programming Formulation

An alternative approach to the least sum of absolute deviations problem may be obtained by defining the positive and negative parts of the ith deviation e_i by setting $f_i = e_i$ and $g_i = 0$ if $e_i \geq 0$, and setting $f_i = 0$ and $g_i = -e_i$ if $e_i < 0$. With these definitions, the least sum of absolute deviations problem may be written in the form:

Minimise

$$\sum f_i + \sum g_i$$

subject to

$$y_i = a + bx_i + f_i - g_i \text{ for } i = 1, 2, \ldots, n$$

$$f_i \geq 0, g_i \geq 0 \text{ for } i = 1, 2, \ldots, n$$

This statement of the problem clearly takes the form of a linear programming problem which may therefore be solved by means of the standard simplex procedure. We shall not discuss this approach to the problem in this book. However, it is interesting to note that the distinction between this solution and the methods of the last two subsections may be illustrated by plotting the successive values of a and b determined by the various procedures in the grid of lines in the ba-plane illustrated in Figures 5.13–5.15.

In his fundamental work in this area, Edgeworth was particularly concerned with the non-uniqueness of the values of the parameters defining the optimal solution. This feature is of less concern nowadays as it is well known that the least sum of absolute deviations problem may be formulated as a linear programming problem, and that for certain data sets such problems have non-unique solutions. The simplest of Edgeworth's resolutions of this difficulty is to choose the centroid of the line segment, plane area, or higher dimensional object defining the set of optimal solutions.

5.4.6   Applications to Voting in Committees

Edgeworth's method of medians has a simple application in an elementary model of voting behaviour. We suppose that there are n individuals on a voting panel, and that these n individuals are required to choose a value for a single parameter, say the managing director's salary. We suppose that the values of this parameter may be naturally ordered along an axis, and that each voter has a preference ordering which is single humped. Then, as the value of the parameter is gradually increased, the voter's level of satisfaction increases (either gradually or by discrete jumps) until his or her level of satisfaction reaches its maximum. Further increases in the value of the parameter above this critical level will lead to successive diminutions in the voter's level of satisfaction, see Figure 5.16.

In this context, each of the members of the panel will vote for an increase in the value of the parameter if it currently lies below his or her preferred value, and for a reduction if it lies above this preferred value. Suppose that there are an odd number of voters on the panel, then the value of the parameter will be increased if the preferred values of more than half of the members are larger than this value. Similarly, a majority of the members of the panel will vote for a reduction in the current value of the parameter if more than half of them have preferred values which lie below this value. As a consequence of this simple analysis, the optimal value of the parameter chosen by the committee as a whole will correspond to the median of the voters' preferred values.

This analysis may easily be related to the discussion of Section 5.1 as the potential energy function corresponding to the ith direct observation has a slope of minus one for values of the parameter lying below that of the ith observation, and a slope of plus one for values lying above this critical value. The potential energy function of the ith observation may therefore be interpreted as a voting function. This function has a slope of minus one if the ith voter

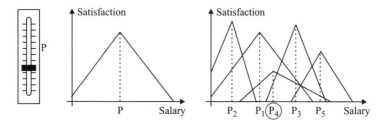

Figure 5.16 Set of single humped preference maps.

prefers a larger value for the parameter than that currently chosen, and a slope of plus one if he or she prefers a smaller value. The sum of these preference functions records the difference between the number of members voting for a decrease in the current value of the parameter over those voting for an increase, see Figure 5.16 above.

The voting problem becomes very complicated when the committee has to choose appropriate values for two parameters, but a special case of this problem has a solution that is closely related to the approximate solution procedure of Subsection 5.4.2. Suppose that there are an odd number of voters who are required to determine values for two parameters. Further, suppose that these voters agree to treat the value of one of the parameters as fixed when determining an appropriate value for the second, and to treat the second parameter as fixed when choosing a value for the first. This iterative procedure continues until the panel reaches an agreed decision, or finds that no such decision is possible. We begin by fixing the value of the second parameter b, then the panel's choice of an optimal value for the first, a, is determined by the middlemost of the n preferred values. Thus, we have to plot the lines of most preferred values of a conditional on b for each of the n committee members, and determine the line of a-medians conditional on the value of b. Similarly, we have to plot a distinct set of most preferred values of b conditional on a for each of the n panel members to determine the line of b-medians conditional on a. The iteration of the committee's deliberations is then determined by plotting the line of a-medians and the line of b-medians in the same diagram. Given an arbitrary starting value for b, we have to find the corresponding point on the line of a-medians. Then, given this value for a, we identify the corresponding value of b on the line of b-medians. This iterative process continues until it is found to converge to a single point, or to a sequence of points in a limit cycle, or it is found to diverge. The process clearly converges to a single point in the case illustrated in Figure 5.17 and to diverge in the case illustrated in Figure 5.18.

For obvious reasons, the solution of difference equations of this type are known as 'cobweb models' in economics. Goldberg (1958) gives a detailed discussion of such models.

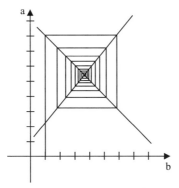

Figure 5.17 Convergent iterative sequence.

5.5 Elemental Set Characterisation of Solutions to Fitting Problems

5.5.1 Elemental Set Solutions

If we are given a set of n equations in q unknowns where n is strictly greater than q, then a natural method of determining values for the unknown parameters of the problem is to choose a set of q equations, and to solve this set of equations for the q unknowns.

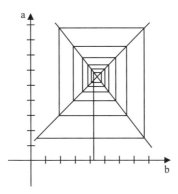

Figure 5.18 Divergent iterative sequence.

This is an extremely ancient method for resolving the general curve fitting problem. It was explicitly employed by Tobias Mayer in 1750, but had been used implicitly by leading practical scientists for more than two thousand years before that date, notably by the early Chinese, Greek, and Islamic astronomers.

In this section, we are concerned with the characterisation of estimation procedures in terms of elemental set solutions of this type. We have already seen that the least sum of absolute deviations problem is characterised by the fact that its solution corresponds to a set of q zero deviations or, failing this, to a linear combination of such solutions.

In the two-dimensional case, each of the lines in a plot of lines corresponds to the values of a and b associated with the zero value of a particular deviation. Thus, the intersection of two such lines identifies an elemental solution to the problem. Identifying all such solutions, we may determine a set of extreme lines such that all elemental solutions to the problem lie on the same side of each of these lines. The successive line segments obtained in this way define the boundary of the convex hull of the set of elemental solutions to the problem, see Figure 5.19.

As an essential preliminary to our exposition, we now have to explain how to construct the convex hull of a set of n points in q-

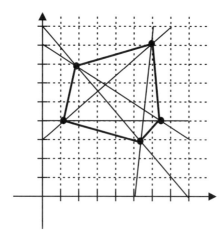

Figure 5.19 Convex hull of a set of points in parameter space.

dimensional parameter space when $n > 2$. Arbitrarily selecting q of these n points, then, in general, we may pass a unique $(q - 1)$-dimensional hyperplane through these q points. If the remaining $n - q$ points lie on one side of this hyperplane and none on the other side, then the $(q - 1)$-dimensional figure with vertices at the q selected points will form part of the boundary of the convex hull of the n points representing the data set. Collecting together all such bounding hyperplanes, we may define a q-dimensional figure with sections of these hyperplanes as boundaries which encloses all n points. This figure is the geometrical representation of the convex hull of the data. Any point lying inside this figure, or on the boundary of it, is a nonnegative linear combination of the given points and no point lying outside it can be expressed in this way.

As stated above, the least absolute deviations solution must lie in this set, and usually on the boundary of it. Our problem in this section is to consider the corresponding characterisation of solutions to the least squares and L_p-norm problems which minimise the weighted or unweighted sum of powers of these deviations.

5.5.2 Elemental Set Characterisation of the Weighted Least Squares Solution

Gilstein and Leamer (1983) have established a simple characterisation of solutions to the general weighted least squares problem. In the two-dimensional case, they note that any pair of values of a and b represents a possible solution to the weighted least squares problem if, and only if, it is possible to find a set of n moduli m_1, $m_2, \ldots, m_n$ satisfying the conditions

$$\sum m_i e_i = 0$$

and

$$\sum m_i x_i e_i = 0$$

where $e_i = y_i - a - bx_i$.

We reformulate this problem by setting $s_i = +1$ and $f_i = m_i e_i$ if e_i is positive; $s_i = -1$ and $f_i = -m_i e_i$ if e_i is negative; and $s_i = 0$ and $f_i = m_i$ if e_i is zero. In this context, we have $m_i e_i = s_i f_i$ and our

problem is to find a set of n strictly positive elements $f_1, f_2, \ldots, f_n$
satisfying the conditions

$$\sum s_i f_i = 0$$

and

$$\sum s_i x_i f_i = 0$$

Implicitly supplying an arbitrary linear function of f_i as the
function to be minimised, this pair of equations clearly takes the
form of a linear programming problem without an explicit objective function. Further, this formulation of the problem indicates
that the existence or nonexistence of a weighted least squares solution depends solely on the existence or nonexistence of a solution
to the problem for the relevant choice of a set of signs $s_1, s_2, \ldots, s_n$.

In this context, Gilstein and Leamer (1983) have shown that
any pair of values a and b will represent a solution to a weighted
least squares problem provided that the corresponding choice of
$s_1, s_2, \ldots, s_n$ represents a bounded region in parameter space. In
particular, if there are $n = 3$ observations then the bounded region
takes the form of a triangle as illustrated by the shaded area in
Figure 5.20. Clearly, the elements a and b cannot take infinite
values, so that weighted least squares estimates cannot correspond
to choices of $s_1, s_2, \ldots, s_n$ which imply that they can.

5.5.3   Explicit Characterisation of the Weighted Least Squares Solution in the Two-Dimensional Case

The above analysis may be generalised to any number of explanatory variables. But, in the two-dimensional case, it is possible to
describe the set of solutions in simple geometrical terms. In this
case, we have to plot the n equations

$$a = y_i - x_i b \qquad i = 1, 2, \ldots, n$$

as lines in the ba-plane. Each of these lines divides the plane into
two distinct regions, so that the full set of n equations will divide

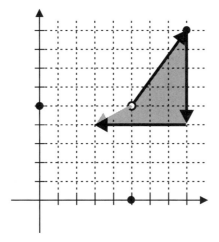

Figure 5.20   Triangular region bounded by three lines in parameter space.

the plane into as many as 2^n regions. Thus, we have only to determine whether or not any particular region is bounded to determine whether the values of b and a lying in this region are possible values for the weighted least squares estimates. The set of all such estimates may readily be obtained from the plot of lines by excluding any region which is not entirely enclosed by a sequence of line segments, again see Figure 5.20.

More formally, Gilstein and Leamer (1983) have shown that the boundary of this set of all possible weighted least squares estimates is easily identified by arranging the equations in order of decreasing slope. Starting at the point in the plane defined by the intersection of the lines with the most positive and the most negative slopes, we proceed along the line with the most positive slope until it intersects with the line with the second most positive slope. We then switch to the line with the second most positive slope, and proceed along this line until it intersects with the line with the third most positive slope. We continue in this way until we reach the line with the most negative slope and return to our starting position. This sequence of lines defines the boundary of the set of weighted least squares estimates. It only remains to determine on a case-by-case basis whether points on the line segments making up the boundary of this set are valid or invalid weighted least squares estimates.

In the particular case of the unweighted least squares estimate, an explicit algebraic form of this characterisation has been established repeatedly by Jacobi in 1841, Glaisher in 1879, Subrahmanyam in 1972, and Ben-Tal and Teboulle in 1990, amongst others, namely that the individual elemental estimators must be weighted by the determinants of the corresponding product matrices and scaled by the sum of these determinants.

5.5.4   The L_p-norm Criterion

Ben-Tal and Teboulle (1990) have extended Gilstein and Leamer's result to a more general class of fitting problems which includes the L_p-norm and minimax problems. Given a positive value for the parameter p, the L_p-norm absolute deviation problem chooses values of a and b to minimise the pth root of the sum of the pth powers of the absolute deviations

$$\left[\sum |y_i - a = bx_i|^P \right]^{\frac{1}{p}}$$

where the pth root operator in this expression is an essential part of the definition of a norm but is only necessary for our purposes in the limit as p tends to infinity. In this limiting case it may readily be shown that the expression is dominated by the largest absolute deviation and is thus identical to the minimax procedure of Chapter 6 which chooses a and b to minimise the largest in absolute value of the deviations, $\max |e_i|$.

If p is finite, then we may delete the pth root operator from this expression and choose a and b to minimise the sum of the pth powers of the absolute deviations:

$$\sum |y_i - a - bx_i|^p$$

Ben-Tal and Teboulle (1990) have shown that, for a class of strictly isotone functions including the L_p-norm function for $0 < p < \infty$,

every solution to the problem which chooses a and b to minimise this function of the deviations lies in the convex hull of
the elemental solutions. Further, for a class of isotone (but not
strictly isotone) functions including the hth largest absolute
deviation functions for $h = 1, 2, \ldots, n$, these authors have
shown that at least one of the solutions to the problem must
lie within the convex hull of the elemental solutions. It is not
necessary for our purposes to know the precise meaning of the
terms *isotone* and *strictly isotone*; we therefore refer interested
readers to the article by Ben-Tal and Teboulle (1990) for definitions.

Restricting these general results to the problems of interest in our Chapters 4 to 7, we find that the solutions to the
least sum of absolute deviations ($p = 1$) and least sum of
squared deviations ($p = 2$) problems necessarily lie in the convex hull of the elemental solutions, whereas the solutions of
the minimax absolute deviations ($h = 1$) and least median of
(squared) absolute deviations ($h = (n + 1)/2$) problems will
necessarily be members of the convex hull only if they are
unique.

5.5.5 Elemental Set Approximations

Numerous robust alternatives to the method of least squares are to
be found in the statistical literature, but all of these procedures
share the common defect that their exact implementation is
extremely burdensome. Rousseeuw and Leroy (1987) have
suggested a possible solution to this difficulty in the specific
context of Rousseeuw's (1984) least median of squares (*LMS*)
procedure. They argue that a sufficiently accurate approximation
to the exact *LMS* solution may be obtained by evaluating the *LMS*
criterion function for a sufficiently large sample of the elemental
set determinations. Hawkins (1993) has shown that this approximate technique yields satisfactory results when applied to a wide
class of robust fitting procedures; in this context, also see
Stromberg (1993).

5.6   Prediction Space Representation

5.6.1   Prediction Space Representation

In Chapter 4, we have presented an observation space representation of the linear fitting problem. In this representation, each of the n observations on the $q + 1$ variables is represented as a point in a $(q + 1)$-dimensional space of observations. We shall now briefly discuss an alternative representation in which each of the sets of n observations on each of the $q + 1$ variables is represented by a point in n-dimensional space.

For simplicity, we shall set $n = 2$ and mark the axes of the two-dimensional plane by y_1 and y_2 respectively. Given any values for the first two observations on the dependent variable, we may locate a point representing these observations at the appropriate position in the horizontal plane defined by these axes. Now, consider any other point (z_1, z_2) on the plane corresponding to the fitted values of these observations, then the distance between their first coordinates is given by $|y_1 - z_1|$ and the distance between their second coordinates by $|y_2 - z_2|$. Combining these two expressions in the component wise L_p-norm measure of distance, we find that the distance between these two points is given by

$$[|y_1 - z_1|^p + |y_2 - z_2|^p]^{\frac{1}{p}}$$

In particular, when $p = 1$ we find that the component wise L_1-norm measure of distance is given by

$$|y_1 - z_1| + |y_2 - z_2|$$

Similarly, when $p = 2$ we have the L_2-norm or Euclidean measure of distance

$$[|y_1 - z_1|^2 + |y_2 - z_2|^2]^{\frac{1}{2}}$$

And when $p = \infty$ we have the L_∞-norm measure of distance

$$\max(|y_1 - z_1|, |y_2 - z_2|).$$

Setting each of these expressions in turn equal to a fixed positive value, say unity, we may identify the set of points (z_1, z_2) at the chosen distance from the given point (y_1, y_2). In the case when p is set equal to 1 we find that the equation

$$|y_1 - z_1| + |y_2 - z_2| = 1$$

defines an oblique square with corners at unit distance due south, north, west, and east of the given point, see Figure 5.21. By definition, all points on the oblique square are at unit distance from the given point, and further contours of this distance function may be obtained by scale transformations of the figure.

In a similar fashion, when p is set equal to 2, the equation

$$|y_1 - z_1|^2 + |y_2 - z_2|^2 = 1$$

defines a circle of unit radius about the given point, see Figure 5.22. Once again, a complete family of contours of this distance function may be obtained by scale transformations of this typical contour.

Similarly, when p is et equal to infinity, the equation

$$\max(|y_1 - z_1|, |y_2 - z_2|) = 1$$

defines a square with sides parallel to the given axes at unit distances due south, north, west, and east of the given point, see Figure 5.23. Again, a complete family of contours may be obtained by scale transformations of the basic figure.

This sequence of diagrams indicate the form of a typical contour for three of the infinite number of L_p-norm measures of dis-

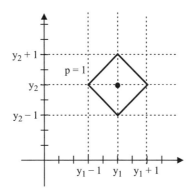

Figure 5.21 Typical contour of the L_1-norm.

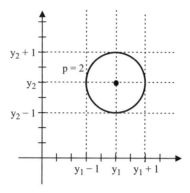

Figure 5.22   Typical contour of the L_2-norm.

tance. In Figure 5.24 we illustrate the relationship between these typical contours for a range of values of p. For values of p less than 1, the sides of the oblique square curve inwards, then, as the value of p increases, the sides straighten out until with $p = 1$ we have the oblique square shown in Figure 5.21. For values of p greater than 1, the same sides tend to curve outwards until with $p = 2$ we have the circle shown in Figure 5.22. For yet larger values of p, the curve near the fixed axial points become straighter until in the limit the typical contour takes the form of a regular square as shown in Figure 5.23.

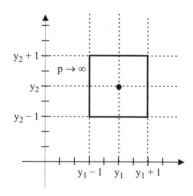

Figure 5.23   Typical contour of the L_∞-norm.

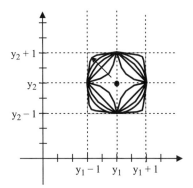

Figure 5.24   Typical contour of the L_p-norm for selected values of p.

In our description of this method of representing data sets, we are largely restricted to the two-dimensional case but, in the present instance, we are able to identify the shape of the typical contour in three dimensions. If $p = 1$ then the oblique square of Figure 5.21 is replaced by an octahedron with eight triangular faces, as illustrated in the interior of Figure 5.25.

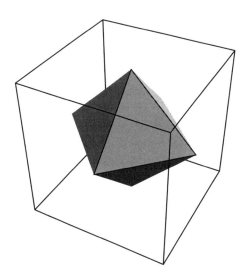

Figure 5.25   Typical contour of the L_1-norm in three dimensions.

If $p = 2$ then the circle of Figure 5.22 becomes a sphere, and if $p = \infty$ then the regular square of Figure 5.23 becomes a cube, as illustrated by the frame of Figure 5.25.

Readers may also be interested to learn the shape of the typical three-dimensional contour for the least median of squared deviations problem of Chapter 7.

This typical contour is most easily constructed by supposing that each of the three pairs of opposite faces of the cube framing Figure 5.25 defines a section from a cylinder with square cross-section and infinite length. This three-dimensional contour and the corresponding one for the two-dimensional case are illustrated in Figures 5.25 and 5.27.

5.6.2 The One-Dimensional Fitting Problem

Using these systems of contour lines, we may readily determine the location of the optimal measure of central tendency corresponding to a chosen measure of distance. Suppose that we have two observations y_1 and y_2, then our problem is to locate a point (z_1, z_2) satisfying $z_1 = z_2$ in such a way as to minimise the distance between these two points. That is, we have to choose a value for the parameter a in such a way that the point (y_1, y_2) is as close as possible to the selected point $(z_1, z_2) = (a, a)$. This problem is most

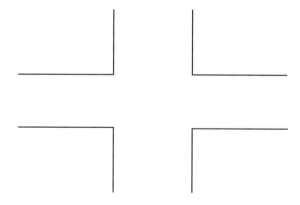

Figure 5.26 Typical contour of the *LMS* criterion in two dimensions.

Figure 5.27   Typical contour of the *LMS* criterion in three dimensions.

conveniently solved by drawing a line of possible positions of the
point (a, a) on a plot of the system of contours representing the
level surfaces of the chosen measure of distance from the point
(y_1, y_2), see Figure 5.28.

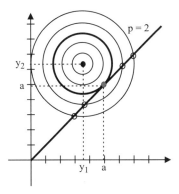

Figure 5.28   Prediction space representation of the least squares fitting
problem.

Now, points of the form (a, a) necessarily lie on a straight line which passes through the origin $(0, 0)$ and the point $(1, 1)$. Plotting this 45 degree line through the origin on a system of circular contours, it is readily seen that the error level associated with certain contours is so low that they do not intersect the line, whilst others associated with larger error levels intersect the line in two places. Our object is to identify the contour associated with the smallest error level that intersects the line through the origin. By comparing the relative curvature of the two functions, it is clear that there is a single solution to this problem, and that this solution corresponds to the point at which the 45 degree line is tangential to a circular contour. Further, as this solution minimises the sum of the squared deviations, this value corresponds to the arithmetic mean of the given observations.

In a similar way, if we draw the line through the origin and the point $(1, 1)$ on the maps of a set of oblique or regular squares, as illustrated in Figures 5.29a and 5.29b, then we may readily identify the contour corresponding to the lowest value of the error function. But, the solutions to these problems may not be unique if one edge of the contour runs along the given line. Thus, the solution of the L_1-norm problem will not be unique in the case illustrated in Figure 5.29a; any value lying between y_1 and y_2 (inclusive) will serve as the median of the given observations. On the other

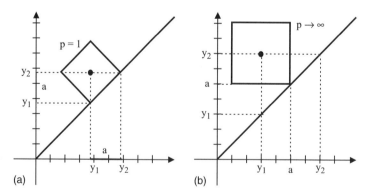

Figure 5.29 Prediction space representation of (a) the least absolute deviations fitting problem, and (b) the minimax absolute deviation fitting problem.

hand, the solution of the L_∞-norm problem illustrated in Figure 5.29b is uniquely determined at the midrange $a = (y_1 + y_2)/2$.

5.6.3 Higher Dimensional Models

In Subsection 5.6.2 we have discussed the case in which the $n = 2$ observations satisfy the condition $z_1 = z_2$. If $n = 2$ but $z_1 \neq z_2$ then we have to draw a line with a nonunit slope in these diagrams; but, with this slight modification, the subsequent analysis takes much the same general form as for the case with $z_1 = z_2$.

In higher dimensions, the observation on the dependent variable is still represented as a single point in n-dimensional space, but the fitted relationship is now represented by a plane or hyperplane of q dimensions which passes through the origin and the q points determined by the observations on the $q < n$ explanatory variables. From this description it is clear that these diagrams are only of symbolic interest when the number of observations exceeds three. It therefore remains to identify the contours of the fitted surface when there are three observations and no more than two explanatory variables. In this context, we have to interpret Figures 5.28, 5.29a, and 5.29b as representing a sequence of concentric spheres, octahedra, or cubes in three-dimensional space intersected by a line or a plane. With these modifications, the argument leading to the optimal solution of the fitting problem again runs much as before.

Thus far in this section, we have assumed that we have a single point in n-dimensional prediction space which we wish to approximate as closely as possible by a single point on a q-dimensional plane or hyperplane in the same space. This problem is readily generalised to the case in which we have a set of $m \geq 2$ points in n-dimensional space which are to be approximated as closely as possible by a set of m points lying on the given q-dimensional hyperplane. The result is unaffected by the scales associated with the individual measures of distance provided that they are independent of each other. In principle, we may restrict the fitted points to a particular region of the given q-dimensional hyperplane. But, if so, then we must specify the relationship between the m measures of distance used in the analysis.

This geometrical model clearly corresponds to the familiar multiple regression problem in which we obtain m sets of q parameter estimates by the joint fitting of a set of n observations on m dependent variables to a set of n observations on q explanatory variables.

References

Ben-Tal, A. and M. Teboulle (1990), A geometric property of the least squares solution of linear equations, *Linear Algebra and Its Applications* **139**: 165–170; supplemented **180**: 5.

Farebrother, R. W. (1987), The historical development of L_1 and L_∞ estimation procedures 1793–1930, in Y. Dodge (Ed.), *Statistical Data Analysis Based on the L_1-norm and Related Methods*, North-Holland Publishing Company, Amsterdam, 37–63.

Farebrother, R. W. (1988), A simple recursive procedure for the L_1 norm fitting of a straight line, *Applied Statistics* **37**: 457–465.

Farebrother, R. W. (1992a), Geometrical foundations of a class of estimators which minimise sums of Euclidean distances and related quantities, in Y. Dodge (Ed.), *L_1-Statistical Analysis and Related Methods*, North-Holland Publishing Company, Amsterdam, 337–349.

Farebrother, R. W. (1992b), Least squares initial values for the L_1 norm fitting of a straight line, *Applied Statistics* **41**: 627–633.

Farebrother, R. W. (1997), Notes on the early history of elemental set methods, in Y. Dodge (Ed.), *L_1-Statistical Procedures and Related Topics*, Institute of Mathematical Statistics, Hayward, California, 161–170.

Farebrother, R. W. (1999), *Fitting Linear Relationships: A History of the Calculus of Observations 1750–1900*, Springer-Verlag, New York.

Gilstein, C. Z. and E. E. Leamer (1983), The set of weighted regression estimates, *Journal of the American Statistical Association* **78**: 942–948.

Goldberg, S. (1958), *Difference Equations*, John Wiley and Sons, New York.

Hawkins, D. M. (1993), The accuracy of elemental set approximations for regression, *Journal of the American Statistical Association* **88**: 580–589.

Koenker, R. (1987), A comparison of asymptotic testing methods for L_1-regression, in Y. Dodge (Ed.), *Statistical Data Analysis Based on the L_1 Norm and Related Methods*, North- Holland Publishing Company, Amsterdam, 287–295.

Koenker, R. and G. W. Bassett (1978), Regression quantiles, *Econometrica* **46**: 33–50.

Koenker, R. and G. W. Bassett (1982), Tests of Linear Hypotheses and L_1 Estimation, *Econometrica* **50**: 1577–1583.

Rousseeuw, P. J. (1984), Least median of squares regression, *Journal of the American Statistical Association* **79**: 871–880.

Rousseeuw, P. J. and A. M. Leroy (1987), *Robust Regressions and Outlier Detection*, John Wiley and Sons Inc., New York.

Stigler, S. M. (1986), *The History of Statistics: The Measurement of Uncertainty before 1900*, Harvard University Press, Cambridge, Massachusetts.

Stromberg, A. J. (1993), Computing the exact value of the least median of squares estimate and stability diagnostics in multiple linear regression, *SIAM Journal of Scientific Computing* **14**: 1289–1299.

Subrahmanyam, M. (1972), A property of simple least squares estimates, *Sankhya (Series B)*, **34**: 355–356.

CHAPTER 6

Minimax Absolute Deviation Method

6.1 One- and Two-Dimensional Means

6.1.1 Observation Space Representation

In this chapter, we consider the fitting problems of earlier chapters but employ the largest absolute deviation as a fitting criterion in place of the sum of squared deviations and the sum of absolute deviations criteria. Our first problem is to adjust the discussion of methods for determining the median and mediancentre of a set of observations to the needs of the minimax absolute deviation criterion. In the one-dimensional case we are given a set of n observations $y_1, y_2, \ldots, y_n$ on a single variable Y and our problem is to choose a value for the parameter a in such a way that the largest in absolute value of the distances $|y_1 - a|, |y_2 - a|, \ldots, |y_n - a|$ between the n observations and the fitted value is minimised. Plotting these observations on the horizontal y-axis we find that we have to identify an interval of minimal length which covers all n observations, see Figure 6.1a. The midpoint of this optimal interval defines the required value of a and the distance from this midpoint to either end of the interval determines the maximum value of the distance function $|y_i - a|$.

This mathematical statement of the problem is readily given a physical form: the endpoints of an arbitrary interval containing all of the observations are marked by the ends of a pair of callipers (or pincers) which are then adjusted in such a way that the interval with these endpoints contains all the observations and has minimal length. The relevant choice of a is then determined by the midpoint of this interval (known as the midrange).

Similarly, in the two-dimensional case we are given a set of n observations (x_i, y_i), $i = 1, 2, \ldots, n$ on two variables X and Y and our problem is to choose values for the parameters c and a in such a way that the largest in absolute value of the Euclidean distances

$$\sqrt{[(x_i - c)^2 + (y_i - a)^2]}$$

is minimised. Plotting these observations in the xy-plane, we find that we have to identify a circle of minimal radius which covers all the observations, see Figure 6.1b. The centre of this optimal circle defines the required values of c and a whilst the radius of this circle determines the maximum distance between an observation and the fitted value.

Once again, this mathematical statement of the problem is readily given a physical form: we suppose that the arbitrary circle is drawn using a pair of compasses whose setting is then adjusted in such a way that the circle is of minimal radius. The relevant choice of c and a are determined by this circle.

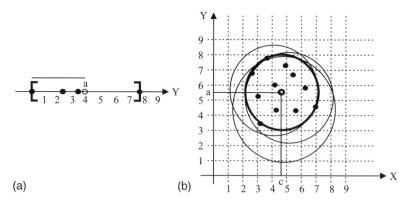

(a)   (b)

Figure 6.1   (a) An interval and (b) a circle containing all the observations.

6.1.2   Mechanical Model Based on Strings and Blocks

For a mechanical model of the minimax absolute deviation problem, we have to combine the strings under tension from the models of Chapter 3 with the pair of horizontal planes from the fundamental model of Chapter 4. We therefore suppose that we are given a pair of horizontal planes, the upper being fixed in position whilst the lower is movable. Holes are drilled through the upper horizontal plane at the n points indicated by the observations on the variables X and Y. A piece of string of given length is passed through each of these holes and attached to the corresponding point on the lower horizontal plane. The upper ends of these strings are tied to a ring lying at an arbitrary point in the upper plane. As the lower (movable) plane is gradually lowered, the strings tighten until, in the limit, it is not possible to lower this plane any further, see Figures 6.2a and 6.2b. This situation clearly determines the position of the ring which minimises the largest of the absolute deviations between the ring and the n holes. Points associated with slack strings are clearly nearer to the ring than those associated with taut strings, and all points associated with taut strings will be at the same distance from the ring. The maximum deviation is thus given by the common length of the taut strings lying in the upper plane.

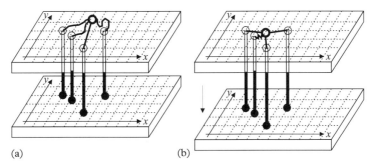

(a)   (b)

Figure 6.2   Mechanical model (a) with slack strings, and (b) with three taut strings.

6.1.3   Parameter Space Representation

As mentioned in Chapter 3, the potential energy function associated with the single observation $y = y_i$ on the single variable Y
takes the form of a V-shaped function

$$P_i = |y_i - a|$$

In the present context, we have a set of n such functions associated
with the n distinct observations on the variable Y. Superimposing
these n V-shaped functions on one another, we have a diagram of
the type indicated in Figure 6.3. The upper bound of this set of
functions represents the maximum distance function for the given
set of n observations. Thus, our problem is to choose a value for
the parameter a in such a way as to minimise the corresponding
value of this upper bound function.

Again, as mentioned in Chapter 3, the V-shaped functions
appropriate for a set of one-dimensional data have to be replaced
by circular cones with V-shaped cross-sections in all directions
when the data relate to the values of two simultaneously observed
variables. Thus, we have to construct a cone of this type at each of
the n points in the horizontal parameter plane and determine their
upper bound to form the overall optimality criterion. If necessary,
this system of cones may be represented by the corresponding
system of circular contours. In this context, we have to combine
the n individual systems of contours and obtain one representing
the largest deviation. Once again, our object is to identify the
parameter values corresponding to the lowest point on this plot.

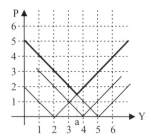

Figure 6.3   Set of V-shaped potential energy functions.

6.2   Simple Linear Regression

6.2.1   Geometrical Solution in Observation Space

In this section, we are once again concerned with the problem of fitting a linear relationship of the form $y = a + bx$ to a set of n observations on the variables X and Y. In this context, we have to choose values for the unknown parameters a and b in such a way as to minimise the largest in absolute value of the deviations from the fitted relationship

$$e_i = y_i - a - bx_i \quad i = 1, 2, \ldots, n$$

Following the discussion of observation space models in earlier chapters, we represent the ith observation on the variables X and Y as a point with coordinates (x_i, y_i) in the two-dimensional Cartesian plane, and the proposed linear relationship between these variables as a straight line with intercept a and slope b in the same Cartesian plane.

In this observation space representation, the problem becomes one of choosing values for the intercept and slope parameters of the arbitrary line in such a way as to minimise the distance (measured parallel to the y-axis) between the line and the point which is furthest from it. A direct implementation of this problem is indicated in Figure 6.4.

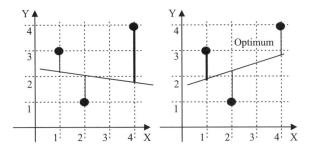

Figure 6.4   Observation space representation of the minimax fitting problem.

An alternative realisation of the same problem may be obtained by bounding the arbitrary line by a pair of parallel lines in such a way that one of the lines passes through the point that is furthest to the north of the arbitrary line whilst the other passes through the point that is furthest to the south of this line. This pair of parallel lines is clearly constructed in such a way that they contain all the observations and pass through at least two of them. If the distance from the arbitrary line to the furthermost point in one direction is not the same as that to the furthermost point in the opposite direction, then the value of the maximum distance function may be reduced by moving the arbitrary line parallel to itself to a position which lies exactly half way between the two bounding lines. We may therefore temporarily forget the fitted line and concentrate on the position of the pair of bounding lines.

In this simplified form, our problem becomes one of choosing the pair of parallel lines containing all the data points which have the least distance between them. This was the approach adopted by Gaspard Clair Françoise Marie de Prony in 1804 in his geometrical solution to this problem.

Having plotted the observations on X and Y as a system of points in the xy-plane, we arbitrarily select a value for the slope parameter b and construct a pair of parallel lines with the chosen slope which just contains all the observations. Given this definition of the pair of parallel lines, we note that each of these lines must pass through at least one of the data points. If each line passes through only one observation, then the distance between them may be reduced by rotating the pair of lines in a clockwise or an anti-clockwise direction about the selected pair of points whilst continuing to pass through them. This solution procedure is illustrated in Figures 6.5a,b,c,d for the case of the $n = 3$ points $(x, y) = (1, 3), (2, 1)$, and $(4, 3)$. The equations given under the diagrams are those of the associated fitted lines.

When the pair of lines encounter a third point, we evaluate the distance between these lines and determine whether it represents a reduction in the value of the fitting criterion. If so, then we adopt this new point in place of the previous point on the same parallel line and continue to rotate the pair of lines in the same direction

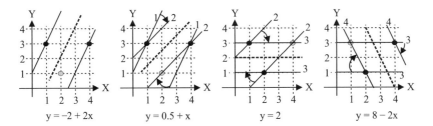

Figure 6.5 Four successive positions of the pair of parallel lines containing all the observations.

about the new pair of points, see Figure 6.5b. We again eventually find a new point and substitute this point for the previous one on the same parallel line. At each stage of this procedure, we have to determine whether the distance between the pair of parallel lines continues to decrease. As soon as it begins to increase, we know that we have found the optimal value of the slope parameter b. If one of the optimal pair of parallel lines passes through the points (x_i, y_i) and (x_j, y_j) whilst the other line passes through the point (x_k, y_k), then the optimal value for the slope parameter is given by

$$b = \frac{(y_j - y_i)}{(x_j - x_i)}$$

and the corresponding value for the intercept parameter by

$$a = \frac{1}{2}[y_j + y_k - b(x_j + x_k)]$$

Further, the absolute values of the functions e_i, e_j, and e_k indicate the distance (measured parallel to the y-axis) between this fitted line and the three or more observations that are furthest from it.

In this context, we may note that the argument underlying De Prony's algorithm implicitly establishes the characterisation of the minimax solution as one determined by a set of three deviations of equal absolute size with two taking one sign and one the opposite sign.

Once again, we may identify a mathematical instrument known as a "pair of parallel rules" which may be used to construct the successive pairs of parallel lines required by the minimax procedure. As its name suggests, this mathematical instrument consists of a pair of parallel rules joined by a pair of parallel cross-ties in such a way as to form an adjustable parallelogram. Until relatively recent times, instruments of this type were employed by astronomers, engineers, and navigators to construct a line that is parallel to a given line and passes through a given point. In this context, the minimax line fitting problem is most easily solved if the observations are marked by nails driven into a horizontal board at the relevant points and the successive pairs of parallel lines determined by means of this instrument.

6.2.2   Influential Observations

If the points plotted in the xy-plane happen to closely follow the path of a straight line, then we may draw a pair of parallel lines near the trend line in such a way that this pair of lines contain all the observations, as indicated in Figure 6.6a. If, now, one of the observations is displaced from its original position by a sufficiently large amount, then one of the parallel lines bounding the set of observations will be determined by the position of this displaced point and one of the extreme points from the original set of observations, whilst the position of the second line will be determined by the other extreme point from the original observations, see Figure 6.6b.

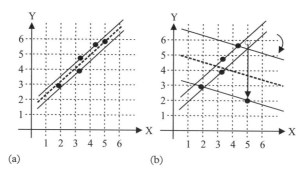

Figure 6.6   Minimax fit of a line (a) when the observations closely follow a trend line, and (b) when one of these observations is displaced.

Thus, the displacement of a single observation may cause the minimax line determined by the bisector of this bounding pair of parallel lines to make an angle of any given size with the general trend of the remaining observations; and, in the limit, it may be constrained to lie at right angles to the general trend. This geometrical demonstration that the minimax line can be extremely sensitive to small changes in the position of one or more of the observations is usually associated with the names of Appa and Smith (1973).

6.2.3 Geometrical Representation in One-Dimensional Parameter Space

The method of Subsection 6.1.3 may readily be generalised to the case in which there is no intercept in the model and the fitted relationship is known to pass through the origin, that is, when $a = 0$ and $y = bx$. The maximum value of the absolute deviation functions may be determined by plotting the n given absolute value functions

$$P_i = |y_i - bx_i| \quad i = 1, 2, \ldots, n$$

in the vertical plane with b on the horizontal axis and P_i on the vertical axis. In this context, the ith V-shaped function has a falling slope of $-|x_i|$ and a rising slope of $+|x_i|$. The function with slopes of -1 and $+1$ discussed in Subsection 6.1.3 is obtained by setting $x_i = 1$.

For our later convenience, we note that the same result may be obtained by combining a plot of the n given functions

$$e_i = y_i - bx_i \quad i = 1, 2, \ldots, n$$

with a second set of n functions embodying their negations

$$e_{n+i} = -y_i + bx_i \quad i = 1, 2, \ldots, n$$

in the vertical plane with b on the horizontal axis and e_i and e_{n+i} on the vertical axis. This alternative construction yields the same diagram if we ignore the portions of the lines below the b-axis, that is, the portions of lines corresponding to negative values of e_i or e_{n+i}.

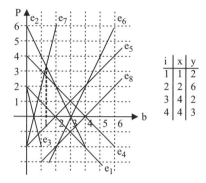

Figure 6.7   Set of absolute value functions and their upper bound.

The line segments and breakpoints of the piecewise linear absolute deviation function are then identified. Further, an inspection of Figure 6.7 will confirm that this optimal solution must correspond to a breakpoint in the piecewise linear upper bound function. Thus, we have only to search this set of breakpoints for the one associated with the smallest value of this function to determine the optimal value of b.

This solution to the minimax problem was developed by Laplace in 1812 without reference to the underlying geometrical structure of the problem. However, this structure is sufficiently close to the surface for us to attribute this geometrical solution to Laplace alone. By contrast with the situation described in Subsection 6.2.5 below, in which practitioners had to wait for more than 160 years for the geometrical counterpart of Laplace's analytical solution to be published, they did not have to wait long for an explicit geometrical interpretation of his analytical procedure in the present context as this was promptly supplied by Augustine-Louis Cauchy and Jean-Baptiste Joséph Fourier.

6.2.4   Geometrical Representation in Two-Dimensional Parameter Space

In their papers of 1824 and 1827, Cauchy and Fourier, respectively, were concerned with the slightly more general problem discussed in Subsection 6.2.1. Their analysis is thus based on the set of $2n$ equations

$$e_i = y_i - a - bx_i \quad i = 1, 2, \ldots, n$$

and

$$e_{n+i} = -y_i + a + bx_i \quad i = 1, 2, \ldots, n$$

Fourier's description of the geometrical form of Laplace's solution of the minimax problem takes the following form. Having plotted the $2n$ functions as oblique planes in three-dimensional Cartesian space, and having discarded the portions of the given planes which lie below the horizontal ba-plane, we find that the uppermost surface of the resulting system of planes defines a mould for a polyhedral vase which is convex towards the horizontal plane.

To find the lowest point of this polyhedral vase, we have to draw a vertical line through any point in the horizontal ba plane, say at the origin, and identify the point at which this line passes through the uppermost oblique plane. Then, we descend this uppermost plane from its point of intersection with the vertical line to a point on an edge of the polyhedral vase, where the selected extreme plane meets a second extreme plane. We then descend this edge until it meets a third extreme plane. Now, three edges meet at this point of intersection, and we arrived at this point by descending one of these edges; we therefore continue our descent along whichever of the unused edges that will take us to the lower point of intersection with a fourth extreme plane. Continuing in this way, we pass from point of intersection to

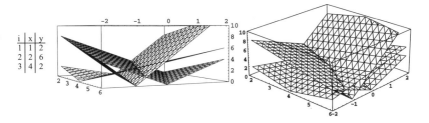

Figure 6.8 Set of absolute value functions in the two parameter case.

point of intersection until we eventually arrive at the lowest point of the polyhedral vase. This lowermost point naturally corresponds to the solution of the minimax absolute deviation problem.

Although this passage gives a clear impression of the geometrical structure of a two-dimensional generalisation of Laplace's solution to the minimax absolute deviations problem, it is still reassuring to have an algebraic description of the proposed solution procedure. Such an algebraic procedure, published by Cauchy in 1824, may be summarised in the following terms. First we have to set a equal to some arbitrary value, say $a = 0$, and to determine the value of b which minimises the maximal deviation. It is easy to show that this optimal value of b must correspond to the equality $e_p = e_q$ of two of the error functions whose b-coefficients x_p and x_q are opposite in sign.

By considering the effect of a small change in the assumed value of a, we may determine whether a should be increased or decreased if the common maximal deviation $e_p = e_q$ is to be reduced. Without loss of generality, we may suppose that a needs to be increased, and choose a third function e_r in such a way that the double equation $e_p = e_q = e_r$, representing the maximal deviation corresponds to the smallest possible increase in the value of a.

Let e_q be the function whose b-coefficient is opposite in sign to that of the incoming function e_r. Then we have to check whether a further increase in the value of a subject to the constraint $e_q = e_r$ will produce a further reduction in the common value of $e_q = e_r$. If so, then we choose a new function e_s in such a way that the double equation $e_q = e_r = e_s$ again corresponds to the smallest possible increase in the value of a. In this way the value of a is gradually increased until it reaches such a value that any further increase would result in an increase in the value of the maximal deviations. In this context, any movement from the values of a and b determined by the final double equation $e_p = e_q = e_r$ will result in an increase in the value of the maximal deviation above the smallest value it is able to attain.

In principle, we may solve this problem by plotting the $n(n-1)/2$ equations of the form $e_i = e_j$ or

$$b = \frac{y_i - y_j}{x_i - x_j}$$

together with the $n(n + 1)/2$ equations of the form $e_i = -e_j$ or

$$a = \frac{1}{2}[y_i + y_j - b(x_i + x_j)]$$

as lines in the ba-plane. (In principle, we should also plot the lines corresponding to the equations $e_i = -e_j$ or $e_i = 0$. But these lines will not be relevant to our analysis unless it is possible to find values for the parameters b and a such that all n equations are satisfied exactly.)

Within this plot of lines, we are interested in the intersection of pairs of lines in which three of the four subscripts are distinct. As in Subsection 5.4.2, the solution of the problem is obtained by moving from one point of intersection of this type to another of the same type in such a way that the corresponding maximal value of the optimality criterion decreases at each stage.

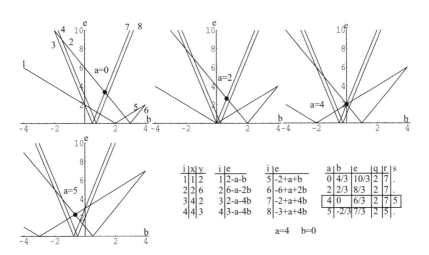

i	x	y
1	1	2
2	2	6
3	4	2
4	4	3

i	e
1	2-a-b
2	6-a-2b
3	2-a-4b
4	3-a-4b

i	e
5	-2+a+b
6	-6+a+2b
7	-2+a+4b
8	-3+a+4b

a	b	e	q	r	s
0	4/3	10/3	2	7	.
2	2/3	8/3	2	7	.
4	0	6/3	2	7	5
5	-2/3	7/3	2	5	.

a=4   b=0

Figure 6.9 Parameter space representation of the minimax line fitting problem.

6.2.5 Simplified Geometrical Solution in Two-Dimensional Parameter Space

The two-dimensional representation of the minimax problem described in the previous section is somewhat involved, not least because the n equations of the problem give rise to a set of $n^2 - n$ lines in the ba-plane. An alternative simpler formulation of the problem is based on a plot of the adjusted deviation functions in the vertical plane, with $e_i + a$ on the vertical axis and b on the horizontal axis, see Figure 6.10. The upper boundary of this plot represents the most positive deviations whilst the lower boundary represents the most negative deviations. Clearly we have to choose a value for b in such a way as to minimise the distance between these two extreme values. For each value of b, the corresponding value of a is obtained as the midpoint of the interval between the two bounding lines.

In view of its early date of 1793, it is somewhat surprising to find that this solution of the minimax problem (also due to Laplace) is based on the minimisation of the largest absolute difference between pairs of deviations

$$\max{}_{i<j} |e_i - e_j|$$

rather than on the direct minimisation of the largest absolute deviation

$$\max{}_i |e_i|$$

Thus, in 1793, and again in 1799, Laplace provided an analytical method for determining the successive line segments and breakpoints of the piecewise linear most positive deviation and most negative deviation functions. Further, he knew from his earlier characterisation of the minimax problem, that its solution must lie at one of these breakpoints, and he had only to search the list of candidate points to identify the optimal values of b and a. Although Laplace's own solution procedure takes an analytical form, see Farebrother (1987, 1999), its geometrical implementation is again sufficiently near to the surface for us to attribute this geometrical solution to Laplace alone. In fact, the first geometrical exposition of this solution to the minimax problem was given by Dolby in 1960 without reference to Laplace's earlier work.

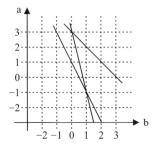

Figure 6.10 Plot of lines in parameter space.

An interesting demonstration of the dual relationship between the solution procedure discussed in Subsection 6.2.1 and that outlined here, is obtained by plotting the lines $a = 3 - b$, $a = 1 - 2b$, and $a = 3 - 4b$ in the ba-plane of Figure 6.10.

These lines correspond to the points $(x, y) = (1, 3), (2, 1)$, and $(4, 3)$ in the xy-plane of Figure 6.5a. Now, plotting the lines $b = 2, b = 1, b = 0$, and $b = -2$ (in that order) parallel to the a-axis, we obtain a sequence of pairs of points on the bounding lines of Figure 6.11 that identify the intercept and slope parameters of the successive pairs of lines employed in Figures 6.5a,b,c,d.

Further, if the midpoints of the successive intervals defined by these lines are connected, then we obtain a piecewise linear function defining the intercept and slope parameters of the fitted lines specified in these diagrams. However, in practice, there is little need

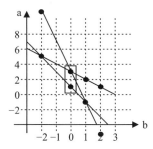

Figure 6.11 Plot of lines identifying successive pairs of parameter values.

to plot this function as the optimal solution may readily be determined without its aid.

6.2.6   Linear Programming Formulation

As in the case of the least sum of absolute deviations problem of Chapter 5, the minimax absolute deviation problem clearly has a linear programming interpretation. Let k be an additional parameter which, with its negation, bounds the value of the ith deviation

$$-k \le e_i \le k \quad i = 1, 2, \ldots, n$$

where the ith deviation is again given by

$$e_i = y_i - a - bx_i \quad i = 1, 2, \ldots, n$$

Then our problem is to choose values for the parameters a and b in such a way as to minimise the value of the additional parameter subject to these constraints. In this context, the minimax problem takes the form of a standard linear programming problem, and may be solved by means of an appropriate variant of the familiar simplex procedure.

Readers familiar with linear programming techniques may find it interesting to represent the successive steps of a standard simplex solution of this problem on the plot of lines in parameter space discussed at the end of Subsection 6.2.4.

For completeness, we note that an alternative approach to the minimax absolute deviation problem with two unknowns is to attempt to identify a set of three equations whose deviations are larger in absolute value than the remaining $n - 3$ deviations. This solution procedure was implemented in an *ad hoc* manner by Laplace in 1786. A determinantal formulation was subsequently developed by Charles de la Vallée Poussin in 1911, see Farebrother (1999). By contrast with Laplace, who apparently proposed to choose successive sets of equations without reference to earlier selections, de la Vallée Poussin introduced an automated selection procedure which Stiefel (1960) subsequently identified with a standard simplex implementation of the linear programming dual formulation of the minimax absolute deviation problem.

6.3 Geometrical Representation of the Harmonic Model

In the special case in which the model to be fitted takes the form of a harmonic relationship

$$y_i = a\cos(\theta_i) + b\sin(\theta_i)$$

we may rewrite the ith equation in the more conventional form

$$y_i \sec(\theta_i) = a + b\tan(\theta_i)$$

and deduce that the distance (measured parallel to the a-axis) from the arbitrary point (b, a) to the ith line is given by

$$f_i = y_i \sec(\theta_i) - a - b\tan(\theta_i)$$

and the corresponding distance measured perpendicular to the ith line by

$$e_i = f_i/\sqrt{(1 + \tan(\theta_i)^2)} = f_i \cos(\theta_i) = y_i - a\cos(\theta_i) - b\sin(\theta_i)$$

Thus, if we draw a circle of radius d in the ba-plane, then all lines which pass through the interior of this circle will correspond to points in the xy-plane which are no more than a distance d from a

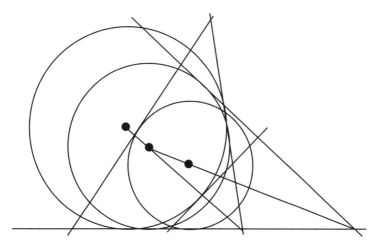

Figure 6.12   Sequence of circles containing all the lines in parameter space.

line whose slope and intercept parameters are given by the centre of the arbitrary circle. Further, those lines which just touch this circle will correspond to points which are at this distance from the fitted line.

Thus, in the *ba*-plane, our problem is to identify a circle of minimum radius that contains all the lines plotted and just touches three (or more) of these lines. Now these three lines are at the same perpendicular distance from the centre of the circle. Thus, in the *ba*-plane, our problem is to identify a circle of minimum radius that contains all the lines plotted and just touches three (or more) of these lines. Now these three lines are at the same perpendicular distance from the centre of the circle. Consequently, the corresponding observations plotted as points in the *xy*-plane are at the same distance measured parallel to the *a*-axis from the line with parameters *a* and *b*. Our problem, therefore, becomes one of finding a circle of minimum radius which contains all of the lines in the *ba*-plane. This realisation of the special case of the minimax problem described in this section is due to Stiefel (1959).

References

Appa, G. and C. Smith (1973), On L_1 and Chebychev estimation, *Mathematical Programming* **5**: 73–87.

Dolby, J. L. (1960), Graphical procedures for fitting the best line to a set of points, *Technometrics* **2**: 477–481.

Farebrother, R. W. (1987), The historical development of L_1 and L_∞ estimation procedures 1793–1930, in Y. Dodge (Ed.), *Statistical Data Analysis Based on the L_1-norm and Related Methods*, North-Holland Publishing Company, Amsterdam, 37–63.

Farebrother, R. W. (1997), The historical development of the linear minimax absolute residual estimation procedure 1786–1960, *Computational Statistics and Data Analysis* **24**: 455–466.

Farebrother, R. W. (1999), *Fitting Linear Relationships: a History of the Calculus of Observations 1750–1900*, Springer-Verlag, New York.

Rousseeuw, P. J. (1984), Least median of squares regression, *Journal of the American Statistical Association* **79**: 871–880.

Stiefel, E. L. (1959), Uber diskrete und lineare Tschebyscheff–Approximationen, *Numerische Mathematik* **1**: 1–28.

Stiefel, E. L. (1960), Note on Jordan elimination, linear programming and Tchebyscheff approximation, *Numerische Mathematik* **2**: 1–17.

Stigler, S. M. (1986), *The History of Statistics: The Measurement of Uncertainty Before 1900*, Harvard University Press, Cambridge, Massachusetts.

CHAPTER 7

Method of Least Median of Squared Deviations

7.1 One- and Two-Dimensional Means

7.1.1 Heuristic Derivation of the Least Median of Squared Deviations Criterion

In this chapter, we consider the same fitting problems as in earlier chapters, except that the median or middlemost absolute deviation is employed as a fitting criterion in place of the sum of the squared deviations, the sum of the absolute deviations, and the largest absolute deviation functions. For simplicity, we shall assume that there are an odd number of observations, as the median absolute deviation is uniquely defined in this context. We shall also assume that the phrase "one-half of the observations" refers to a set of $(n + 1)/2$ of these n observations.

In Chapter 5, we briefly mentioned the problem of choosing the parameters of the model in such a way as to minimise the sum of the pth powers of the absolute differences between the observed values and the fitted values. In particular, we discussed the case $p = 1$ when we are interested in minimising the sum of the absolute deviations and the case $p = 2$ when we are concerned with the corresponding minimisation of the sum of the squared deviations.

As a further generalisation of this principle, we may divide the optimality criterion by the number of observations n and replace the word *sum* by the word *average* in the statement of the fitting problem. This simple modification of the optimality criterion has no effect on the values of the parameters selected. However, we may now generalise the revised optimality principle by replacing the average or arithmetic mean of the pth powers of the absolute deviations by some other measure of central tendency. Following Rousseeuw (1984), we shall pay particular attention to the case in which we are interested in minimising the *median* of the pth powers of the deviations. In this context, we choose values for the parameters of the model in such a way as to minimise the median or middlemost of the n terms $|e_1|^p$, $|e_2|^p$, . . . , $|e_n|^p$ where the value of p is still to be specified. In fact, when n is odd we obtain exactly the same result whichever positive value is selected for the parameter p.

Setting $p = 1$, we find that we are concerned with the problem of minimising the middlemost of the terms $|e_1|$, $|e_2|$, . . . , $|e_n|$. Alternatively, by setting $p = 2$ we find that we have to choose the parameters in such a way as to minimise the middlemost of the terms e_1^2, e_2^2, . . . , e_n^2. This last choice justifies our naming this procedure "the least median of squared deviations procedure." This procedure is now quite widely employed as the resulting parameter estimates are essentially determined by one-half of the observations and are thus robust to gross errors in the unused half of the data set.

7.1.2   One- and Two-Dimensional Means

Our first problem is to adjust the discussion of methods for determining the midrange of a set of observations to the needs of the least median of absolute distances criterion. In the one-dimensional case, we are given a set of n observations $y_1, y_2, . . . , y_n$ on a single variable Y, and our problem is to choose a value for the parameter a in such a way that the middlemost in absolute value of the distances $|y_1 - a|$, $|y_2 - a|$, . . . , $|y_n - a|$ between the n observations and the fitted value is minimised. Plotting these observations on the horizontal y-axis we find that we have to identify an interval of minimal length which covers one-half of the observations. The midpoint of this optimal interval defines the required value of a, and the distance from its midpoint to either end determines the median value of $|y_i - a|$. This situation is illustrated in Figure 7.1 in which the optimal interval covers three of the five observations.

As in Chapter 6, this mathematical statement of the problem is readily given a physical form: the endpoints of an arbitrary interval containing one-half of the observations are marked by the ends of a pair of callipers which are then adjusted in such a way that the interval with these endpoints contains one-half of the observations and has minimal length. The relevant choice of a is then determined by the midpoint of this interval.

Similarly, in the two-dimensional case, we are given a set of n observations (x_i, y_i) $i = 1, 2, \ldots, n$ on two variables X and Y and our problem is to choose values for the parameters c and a in such a way that the middlemost in absolute value of the Euclidean distances

$$\sqrt{[(x_i - c)^2 + (y_i - a)^2]}$$

is minimised. Plotting these observations in the xy-plane we find that we have to identify a circle of minimal radius which covers one-half of the observations. The centre of this optimal circle defines the required values of c and a whilst the radius of this circle determines the median distance between the observations and the fitted value, see Figure 7.2 in which the optimal circle covers six of the eleven observations.

Once again, this mathematical statement of the problem is readily given a physical form: we suppose that the arbitrary circle is drawn using a pair of compasses whose setting is then adjusted in such a way that the circle is of minimal radius. The relevant choice of c and a are determined by the centre of this circle.

Figure 7.1   Interval covering one-half of the observations.

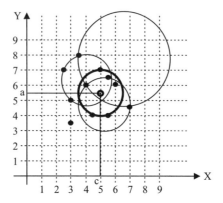

Figure 7.2   Circle covering one-half of the observations.

In principle, it is also possible to generalise the mechanical model of the minimax absolute deviation problem to the present context. Suppose that we are given a pair of horizontal planes at an arbitrary distance apart. Holes are drilled through the upper horizontal plane at the n points indicated by the observations on the variables X and Y. Selecting a set of $(n + 1)/2$ of these holes, we pass a piece of string of given length through each of the selected holes and attach it to the corresponding point on the lower horizontal plane. The upper ends of these $(n + 1)/2$ strings are tied to a ring lying at an arbitrary point in the upper plane. As the lower (movable) plane is gradually lowered, the strings tighten until, in the limit, it is not possible to lower this plane any further. This situation clearly determines the position of the ring which minimises the largest of the absolute deviations between the ring and the $(n + 1)/2$ selected holes. Points associated with slack strings are nearer to the ring than those associated with taut strings and all points associated with taut strings will be at the same distance from the ring. The maximum deviation of the $(n + 1)/2$ selected points is given by the length of the taut strings lying in the upper horizontal plane.

This mechanical model describes the minimisation of the median of absolute deviations criterion for a particular choice of $(n + 1)/2$ observations. In fact, this criterion has to be minimised over all such selections of one-half of the observations.

7.1.3   Parameter Space Representation

As mentioned in Chapter 3, the potential energy function associated with the single observation $y = y_i$ takes the form of a V-shaped function

$$P_i = |y_i - a|$$

In the present context, we have a set of n such functions associated with the n distinct observations on the variable Y. Superimposing these n V-shaped functions on one another, we have a diagram of the type indicated in Figure 7.3.

   As in Chapter 6, the upper bound of this set of functions represents the maximum distance function for the given set of n observations. If this upper bound is removed, we have a new upper bound which represents the second largest absolute distance function, see Figure 7.4.

   Again, removing this upper bound, we have a new upper bound which records the third largest absolute distance, and so on, see Figure 7.5.

   Thus, our problem becomes one of choosing a value for a in such a way as to minimise the value of the $(n + 1)/2$th upper bound function.

   Clearly, the irregular nature of the median of squared deviations function makes it very difficult for us to determine the opti-

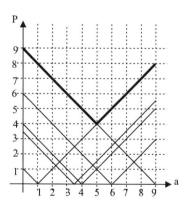

Figure 7.3   Set of V-shaped potential energy functions.

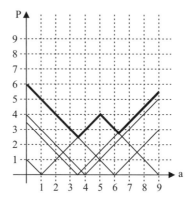

Figure 7.4 Set of potential energy functions with the upper bound removed.

mal value of a. The fitting problem becomes yet more involved in two (and higher) dimensions, when the V-shaped functions appropriate for one-dimensional data have to be replaced by circular cones with V-shaped cross-sections in all directions. In this context, we have to construct a cone of this type at each of the n points in the horizontal parameter plane, identify the relevant portion of the $(n + 1)/2$th upper bound function, before locating the lowermost point on this function.

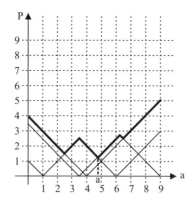

Figure 7.5 Set of potential energy functions with two upper bound removed.

7.2   Simple Linear Regression

7.2.1   Physical Solution in Observation Space

In this section, we are once again concerned with the problem of fitting a linear relationship of the form $y = a + bx$ to a set of n observations on the variables X and Y. In the present context, we have to choose values for the unknown parameters a and b in such a way as to minimise the middlemost in absolute value of the deviations from the fitted relationship

$$e_i = y_i - a - bx_i \quad i = 1, 2, \ldots, n$$

Following the discussion of observation space models in Chapter 4, we represent the ith observation on the variables X and Y as a point with coordinates (x_i, y_i) in the two-dimensional Cartesian plane, and the proposed linear relationship between these two variables as a straight line in the same Cartesian plane.

In this observation space representation, the problem becomes one of choosing values for the intercept and slope parameters of the arbitrary line in such a way as to minimise the distance (measured parallel to the y-axis) between the line and the point which is $(n + 1)/2$th furthest from it.

As explained in Chapter 6, an alternative realisation of the same problem is to bound the arbitrary line symmetrically by a pair of parallel lines in such a way that they contain or pass through one-half of the observations. If one or both of the bounding lines does not pass through one or more points then the value of the median distance function may be reduced by moving the pair of bounding lines to a position in which they both pass through at least one data point.

Thus, we have bounded the arbitrary line by a pair of parallel lines containing one-half of the data points and choose the pair of lines with the least distance between them. This was the approach adopted by Rousseeuw (1984) in his solution of the least median of squares line fitting problem. Unfortunately, there is no simple iterative procedure for determining the optimal choice of the parameters. Once again, we may employ the mathematical instrument known as a "pair of parallel rules" to construct the successive pairs of parallel lines required by this procedure.

7.2.2   Influential Observations

In Subsection 6.2.2 we described an extreme example of the line fitting problem in which the observations closely follow a straight line in the xy-plane, so that the optimal pair of parallel lines containing all n observations closely follow the course of this line. If, however, a few of the observations are removed from their original positions, then the minimax fitted line will be affected to such an extent that, in the limit, it may lie at right angles to the general trend of the remaining observations. By contrast, the least median of squares procedure does not suffer from this defect as almost one-half of the observations may be given arbitrary positions without affecting the location of the fitted line.

7.2.3   Geometrical Representation in Parameter Space

As in Chapter 6, the method of Subsection 7.1.3 may readily be generalised to the case when there is no intercept in the model and the fitted relationship is known to pass through the origin. That is, when $a = 0$ and $y = bx$. The median of the absolute value functions may be determined by plotting the n given absolute value functions

$$P_i = |y_i - bx_i| \quad i = 1, 2, \ldots, n$$

in the vertical plane with b on the horizontal axis and e_i on the vertical axis. In this context, the ith V-shaped function has a falling slope of $-|x_i|$ and a rising slope of $|x_i|$.

For our later convenience, we note that we may obtain the same result by combining a plot of the n given functions

$$e_i = y_i - bx_i \quad i = 1, 2, \ldots, n$$

with a second set of n functions embodying their negations

$$e_{n+i} = -y_i + bx_i \quad i = 1, 2, \ldots, n$$

in the vertical plane with b on the horizontal axis and e_i on the vertical axis. This alternative construction yields the same diagram if we ignore the portions of the lines lying below the b-axis, that is, the portions of lines corresponding to negative values of e_i or e_{n+i}.

The line segments and breakpoints of the successive piecewise linear hth largest absolute deviation functions are then identified for $h = 1, 2, \ldots, (n + 1)/2$. We have only to search the set of breakpoints of the $(n + 1)/2$th largest absolute deviation function to find the one associated with the smallest value of this function and thus determine the optimal value of b.

This procedure may readily be generalised to the problem in two unknowns discussed in Subsection 7.2.1. The analysis is thus based on the $2n$ equations

$$e_i = y_i - a - bx_i \quad i = 1, 2, \ldots, n$$

and

$$e_{n+i} = -y_i + a + bx_i \quad i = 1, 2, \ldots, n$$

Again, we have to peel off the successive upper bounding functions until the relevant one is exposed.

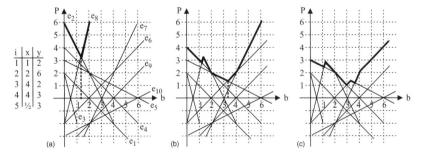

Figure 7.6 (a) A set of potential energy functions in the two parameter case, (b) with one upper bound removed, and (c) with two upper bounds removed.

7.2.4 An Alternative Geometrical Solution

By contrast with the relatively simple geometrical method for determining the values of a and b to minimise the largest absolute deviation, the procedure for determining the corresponding values which minimise the $(n + 1)/2$th largest absolute deviation are relatively complex and have several features in common with the geometrical solution of the least sum of absolute deviations procedure.

Our first scheme (not illustrated) is based on the n equations

$$a = y_i - bx_i \quad i = 1, 2, \ldots, n$$

Plotting these equations as functions of b in the ba-plane, we successively peel off the upper and lower bounds from the plot of lines to reveal the successive bounding lines or strands. Setting $h = (n + 1)/2$ for ease of exposition, we match up the first and hth of these bounding lines, the second and $(h + 1)$th lines,..., and the $(n - h + 1)$th and nth lines; then we have to identify the value of b associated with the minimum distance between any one of these $n - h + 1$ pairs of lines. The corresponding optimal value of a is then obtained as the midpoint of this minimal distance.

7.2.5 A Second Geometrical Solution

In this section, we shall not be concerned with the successive line segments and breakpoints of the piecewise linear most positive deviation and most negative deviation functions. Instead, we shall modify the technique discussed in Subsection 5.4.2 to our present purpose. Figure 7.7 illustrates a plot of $n = 5$ lines in the ba-plane. For each value of b, we may read off the values of the n functions $y_1 - bx_1, y_2 - bx_2, \ldots, y_n - bx_n$. In this context, our problem becomes one of choosing an interval of minimal length which covers $(n + 1)/2$ of these values, and to identify the midpoint of this interval with the value of a. For a range of nearby values of b, the endpoints of this interval will be defined by the same pair of lines. Hence, the set of midpoints defining the relationship between the values of b and a will also take the form of a line segment.

However, by contrast with the results in Subsections 5.4.2 and 6.2.5, the individual line segments will not usually be connected to one another. The required values of a and b are then obtained by searching these line segments to identify the values of b and a which minimise the median absolute distance function.

The fact that the individual line segments are not usually connected implies that in certain circumstances, small changes in the values of the observations may cause the optimal values of the parameters to jump from a point on one of the line segments in Figure 7.7 to a point on another line segment. This is a somewhat disconcerting feature to be exhibited by an estimation technique that is supposed to be robust to small changes in the observed values of X and Y, see Hettmansperger and Sheather (1992) for a worked example. However, these discontinuous changes in the values of the parameters are clearly bounded, and the imposition of continuity on the parameter estimates would restrict us to a class of procedures that are not robust to small changes in the values of the observations.

In passing, we note that we could also have plotted the sequence of line segments identifying the optimal value of b for each value of a, but the result would hardly represent a practical fitting procedure.

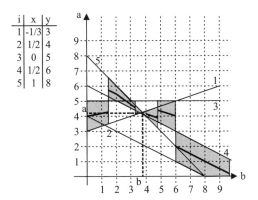

Figure 7.7 Optimal a-intervals for a range of values of b.

7.2.6   Nonlinear Programming

Let z_i be an indicator which takes the value unity if the ith observation lies on, or between, the pair of parallel lines and the value zero if it does not. Further, let k be an additional parameter which, with its negation, bounds the value of the selected deviations

$$-k \leq z_i e_i \leq k \quad i = 1, 2, \ldots, n$$

or, equivalently,

$$-k + (1 - z_i)c \leq e_i \leq k + (1 - z_i)c \quad i = 1, 2, \ldots, n$$

where c is an arbitrarily large constant, and where the ith deviation is again given by

$$e_i = y_i - a - bx_i \quad i = 1, 2, \ldots, n$$

Then our problem is to choose values for the parameters a and b in such a way as to minimise the value of this additional parameter subject to the constraint that at least $(n + 1)/2$ of the indicators take unit values. In this context, this problem takes the form of a mixed integer linear programming problem:

minimise k

subject to
$$-k \leq y_i - a - bx_i+ \leq k$$
and
$$\sum z_i \geq (n + 1)/2$$

   However, the sheer complexity of the least median of squares problem implies that all the computational procedures discussed in this section are likely to be impractical. We are therefore often obliged to employ the simulation procedures proposed by Rousseeuw and Leroy (1987), Hawkins (1993), and Stromberg (1993) discussed in Subsection 5.5.5 in order to obtain good approximations to the least median of squares values.

7.3 Generalisation to Minimum Volume Median Ellipsoids and Ellipsoidal Cylinders

In Section 7.1 we discussed a measure of central tendency defined by the centre of a circle of minimum radius (and therefore minimum area) which contains one-half of the observations. This idea may readily be generalised by defining an ellipse of minimal area which contains one-half of the observations. In three and higher dimensions we associate a measure of central tendency with the centre of a q-dimensional ellipsoid of minimal q-dimensional volume which just contains one-half of the observations. The fundamental advantage of the measure of central tendency associated with the centre of such an ellipse or ellipsoid, is that it is affine equivariant to linear deformations of the observations, see Rousseeuw (1984) for details.

In the three-dimensional case, the measure of central tendency defined by the minimum volume ellipsoid may be generalised to yield a line of best fit associated with the centre of a cylinder with an elliptical cross-section which contains one-half of the observations and a plane of best fit associated with the plane bisector of a pair of parallel planes which again contain one-half of the observations, see Farebrother (1994). Both of these statements clearly represent alternative three-dimensional generalisations of the pair of parallel lines containing one-half of the observations described in Section 7.2.

References

Farebrother, R. W. (1994), On hyperplanes of closest fit, *Statistical Computing and Data Analysis* **19**: 53–58.

Hawkins, D. M. (1993), The accuracy of elemental set approximations for regression, *Journal of the American Statistical Association* **88**: 580–589.

Hettmansperger, T. P. and S. J. Sheather (1992), A cautionary note on the method of least median of squares, *The American Statistician* **46**: 79–83.

Rousseeuw, P. J. (1984), Least median of squares regression, *Journal of the American Statistical Association* **79**: 871–880.

Rousseeuw, P. J. and A. M. Leroy (1987), *Robust Regressions and Outlier Detection*, John Wiley and Sons, New York.

Stromberg, A. J. (1993), Computing the exact value of the last median of squares estimate and stability diagnostics in multiple linear regression, *SIAM Journal of Scientific Computing* **14**: 1289–1299.

CHAPTER 8

Mechanical Models of Metric Graphs

8.1 Introduction

In Chapter 3, we discussed the two-dimensional case of a familiar problem in multivariate analysis which may be expressed in the following geometrical form. We are given a set of n observations on q variables; these n observations may be represented by n points in q-dimensional Euclidean space. Our problem is that of choosing an additional point in such a way that the sum of the absolute (Euclidean) distances from this additional point to the n given points is minimised. This problem defines a measure of central tendency known as the mediancentre or the centre of population.

As a simple generalisation of the mechanical model of Chapter 3, we find that the relevant model for the present problem is obtained by placing an eyelet at each of the n points indicated by the data; n strings with unit weights at their free ends are then passed through these n eyelets before being tied to a ring at an arbitrarily selected point. Assuming that the weights are attracted by a force acting at right angles to the q-dimensional space of observations, we find that the potential energy of the system is proportional to the sum of the absolute distances, and thus that the mediancentre is associated with the minimum value of this function.

In the present chapter, we shall extend this mechanical model to a class of data sets which may be presented in the form of metric graphs. In particular, we shall discuss the analysis of sets of data generated by transitive and nontransitive pairwise preference orderings.

8.2 Metric Graphs

We initiate our discussion by defining the graph theoretic terms we shall need in this chapter. A *graph* consists of a set of *nodes* and a set of *paths* connecting pairs of nodes. A graph becomes a *metric graph* when the lengths of these paths are known. For simplicity, we shall assume that each node is connected to every other node by one or more sequences of paths. Thus, our immediate task is to generalise the mechanical model of Section 8.1 to data sets which may be represented in the form of metric graphs.

As before, we represent the nodes of the graph by points in a Euclidean space of suitable dimension. But, in the present context, we are obliged to make specific provision for the non-Euclidean nature of the distances between pairs of nodes. We therefore represent these distances by flexible hollow tubes of appropriate length. Having constructed this model, we select an arbitrary point in one of the tubes and, for each, node, we pass a weighted string through an eyelet at this node and connect it to a ring located at the

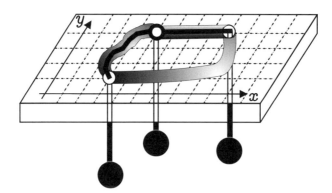

Figure 8.1 Physical representation of a metric graph with three nodes.

arbitrary point by passing the string through the shortest sequence of tubes joining the two points.

The full set of tubes joining the n nodes to the arbitrary point defines a *spanning tree* which connects each point to every other point without cycles, see Figures 8.1 and 8.2. If the ring representing the trial value is now released, the model will locate the position within the selected spanning tree which mini-mises the sum of the path lengths. When this point is in a state of equilibrium, the force tending to full the ring in one direction along a tube must balance the force tending to pull it in the opposite direction.

Thus, as in the case of the median of a set of direct observa-tions, the median point of a given spanning tree is uniquely defined when there are an odd number of observations, see Monjardet (1991). Conversely, if there are an even number of observations and an optimal solution occurs at an interior point of one of the tubes, then any point in this tube together with its end points will also be optimal, see Hakimi (1964).

In the last two paragraphs, we have described a mechanical model for the median of a metric graph which happens to take the form of a spanning tree. In general, we would have to apply this model to each of the trees which span the given metric graph and choose the point which minimises the optimality criterion over all spanning trees.

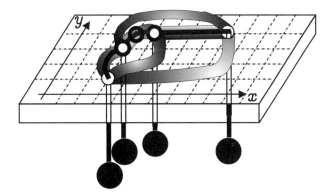

Figure 8.2 Physical representation of a metric graph with four nodes.

For example, if the graph consists of four nodes and five paths arranged in the form illustrated in Figure 8.3, then the ring will be in equilibrium at the point B if all four nodes are connected directly to the ring at this point. Alternatively, the ring will be in equilibrium at any point on the transverse line BD if it is connected directly to the points B and D, to the point A via B, and to the point C via D. Similarly, the ring will be in equilibrium at any point on the line BC if it is connected directly to the points B and C, to the point A via B, and to the point B via C. Further, these three solutions are associated with total path lengths of $AB + BC + BD$, $AB + 2BD + CD$, and $AB + 2BC + CD$ respectively, so that, in principle, the solution which minimises the total path length may be identified by extending our analysis to cover all possible configurations of the problem. Unfortunately, it does not seem possible to automate the choice of an optimal solution to problems of this type. In particular, the optimal solution cannot be found by the successive adjustment of the current choice of a spanning tree. Thus, it is clear that this model does not furnish a suitable basis for a computational procedure. Such procedures are usually based on iteratively adjusted variants of the mechanical model of Section 8.1, see Eilon, Watson-Gandy, and Christofides (1971) and Franksen and Grattan-Guinness (1989, pp. 215–216) for details.

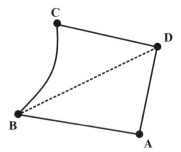

Figure 8.3   Schematic representation of a metric graph with four nodes.

8.3   Pairwise Preference Orderings

In this section, we consider a special case of the problem of Section 8.2, in which a panel of voters is required to combine their individual pairwise preference orderings of a set of candidates into a single overall pairwise preference ordering.

Suppose that there are p candidates under consideration, then, we may denote them by the letters $A, B, C, \ldots$, and deduce that there are $m = p(p-1)/2$ distinct pairs of candidates and thus the same number of lexicographic preference statements of the type "A is preferred to B", "A is preferred to C", "B is preferred to C", ..., or, more briefly, $A > B, A > C, B > C, \ldots$. In this context, each of the individual voters is required to consider whether each of these m lexicographic statements is true or false, and to return his or her decision in the form of a complete pairwise preference ordering to the panel for processing. The problem of incomplete preference orderings may readily be resolved by allowing the individual voters to select weighted combinations of complete preference orderings. For example, if a voter prefers A to B and A to C, but is not sure whether he or she prefers B to C or C to B, then he or she may return the ordering $A > B = C$ by selecting the orderings $A > B > C$ and $A > C > B$ each with weight $1/2$.

In this problem, there are a total of 2^m complete pairwise preference orderings which may most simply be represented as sequences of ones and zeros (corresponding to true and false statements respectively). The resulting binary sequences may then be identified with the vertices of an m-dimensional binary lattice. In this context, it seems natural to adopt as our measure of distance the number of disagreements between the ith individual's binary sequence and that of the trial solution. With this measure of distance, our problem becomes one of choosing a single overall preference ordering in such a way as to minimise the total number of disagreements between the selected ordering and the orderings of the n individual voters. A natural mechanical model is obtained by connecting pairs of adjacent vertices of the lattice by hollow tubes of unit length and the ith individual's selection to a ring at the trial solution by a string under unit tension which passes along a path of minimal length.

As an immediate consequence of this choice of mechanical model and the associated measure of distance, we find that we may resolve the forces acting on the ring parallel to the m axes of the lattice and deduce that this problem is equivalent to one in which the panel arrives at its overall decision by holding a sequence of votes on each of the pairwise comparisons in turn. Further, when there are an odd number of voters on the panel, this model will necessarily yield a majority decision on each of the m lexicographic statements defining the m dimensions of the lattice.

As a simple numerical example of this procedure, we suppose that we have a panel of five voters expressing their preferences in respect of three candidates; and that two voters prefer the ordering $A > B > C$, two the ordering $B > C > A$, and one the ordering $C > A > B$. Then we have to place a weight of two units at the point $(1, 1, 1)$, a weight of two units at the point $(0, 0, 1)$, and a weight of one unit at the point $(1, 0, 0)$, see Figure 8.4. With these assignments, we find that the panel as a whole prefers A to B by three votes to two, it prefers C to A by three votes to two, and it prefers B to C by four votes to one. Thus, the panel's solution to this problem corresponds to the point $(1, 0, 1)$ which denies each of the voters one of his or her preferences and thus yields a distance measure of five units.

Although this model necessarily arrives at a firm decision when there are an odd number of voters, it still suffers from the

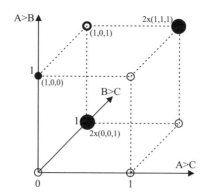

Figure 8.4   Pairwise preference orderings represented by a binary lattice.

defect that the decision it arrives at may involve sets of contra-dictory statements (such as the statements $A > B$, $B > C$, and $C > A$ corresponding to the point $(1, 0, 1)$ in Figure 8.4) even though none were present in the individual voting returns. We shall outline a possible solution to this problem in the following section. However, the consequent loss of the regular structure of the binary lattice means that we are obliged to use a general purpose optimi-sation procedure to determine the consistent pairwise preference ordering which minimises the total number of disagreements between the selected preference ordering and those of the n indi-vidual voters.

8.4   Transitive Preference Orderings

An obvious solution to the consistency problem identified at the end of the previous section is to delete from the lattice all points which correspond to sets of inconsistent statements. For example, when there are three candidates A, B, and C, and thus three lexicographic preference statements $A > B$, $A > C$, and $B > C$, we have to delete two diagonally opposite points $(1, 0, 1)$ and $(0, 1, 0)$ (corresponding to the inconsistent statements $A > B$, $B > C$, $C > A$, and $A > C$, $C > B$, $B > A$) from the three-dimensional lattice to obtain an appropriate geometrical figure. This figure illustrated

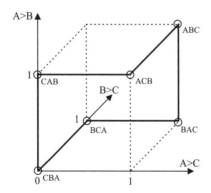

Figure 8.5   A binary lattice with inconsistent vertices removed.

in Figure 8.5, has six sides of unit length that are sequentially connected in such a way that a further slight adjustment to the figure will transform it into a regular plane hexagon, see Figure 8.6.

In this context, we may label the successive vertices of the hexagon with the symbols ABC, ACB, CAB, CBA, BCA, and BAC in accordance with their respective positions in the three-dimensional lattice of Figure 8.5. Here the symbol ABC represents the ordering $A > B > C$ associated with a consistent position on the lattice.

Returning to our numerical example in this context, we find that the combinations $A > B > C$, $B > C > A$, and $C > A > B$, or ABC, BCA, and CAB, are spaced two units apart around the circumference of a hexagon of unit side. It is thus easily deduced that either of the first two combinations will serve as an optimal consistent pairwise preference ordering for the panel as a whole, see Figure 8.7. Further, each of these consistent solutions is associated with a distance measure of six units which is marginally greater than the measure of five units for our earlier inconsistent solution, see Figure 8.7.

This new symbolic notation offers an alternative method for determining the main geometrical features of the $(p-1)$-dimen-

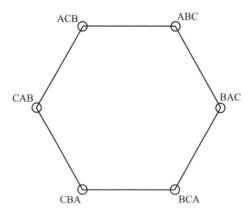

Figure 8.6 Hexagonal representation of consistent pairwise preference orderings.

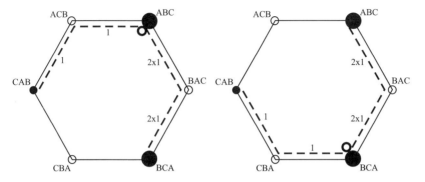

Figure 8.7 Graphical representation of the two solutions to the optimisation problem.

sional figure that is obtained when the points corresponding to inconsistent combinations are deleted from the m-dimensional binary lattice. An examination of this sequence of symbols reveals that the successive terms have a single element in common and that this common element is alternately in the first and last position. We therefore deduce that, at each stage of a complete circuit of the hexagon, a pair of adjacent letters exchange places, a single preference statement is reversed, and the (unidirectional) measure of distance from any given starting position to the present position is increased by unity. This argument is illustrated in Table 8.1, where ABC is taken as the starting position.

Table 8.1   Circuit of the Symbolic Hexagon

Symbol	Lexicographic Statement	Distance
ABC	$A > B, A > C, B > C$	0
ACB	$A > B, A > C, C > B$	1
CAB	$A > B, C > A, C > B$	2
CBA	$B > A, C > A, C > B$	3
BCA	$B > A, C > A, B > C$	4
BAC	$B > A, A > C, B > C$	5
ABC	$A > B, A > C, B > C$	6
ACB	$A > B, A > C, C > B$	7

8.5   Three-Dimensional Models

Now, consider extending this argument to the case of four candidates. The four elements *A*, *B*, *C*, and *D* may appear as the first or the last element in a sequence of four elements. For each of these eight possibilities we have a cycle of six orderings of the remaining three elements. Further, each of these cycles may be represented by a hexagon which has three pairs of elements in common with three other hexagons, see Figure 8.8.

Given the relative positions of any two adjacent hexagons, the sides of the other six may be matched up to generate a symmetrically truncated hollow octahedron. Readers are invited to demonstrate this conclusion for themselves by copying the eight hexagons given in Figure 8.8 and connecting them in the manner indicated in Figure 8.9.

We may further clarify the nature of this geometrical figure by constructing a pyramid on an open square base by joining four equilateral triangles of side three units at their edges. We may then obtain a regular octahedron of side three units by joining two such pyramids at their bases. This regular octahedron may now be truncated in the required manner by removing an equilateral triangle of unit side from each corner of each of the eight equilateral

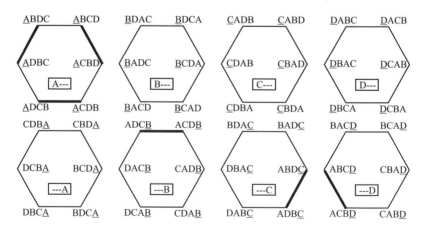

Figure 8.8   Set of eight symbolic hexagons.

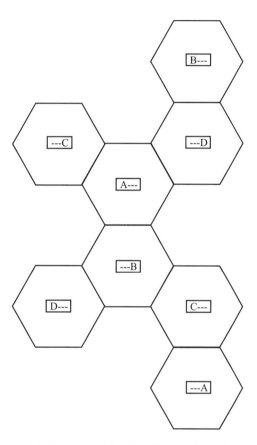

Figure 8.9 The eight hexagons joined to form a plane figure.

triangles that form its faces. Further, if the vacant axial positions of the octahedron are sealed with six unit squares, then the resulting Figure 8.10 is known as a *tetrakaidekahedron*. This is mathematical Greek for a fourteen-faced figure, in much the same way as *tetrahedron* and *octahedron* is mathematical Greek for a four-faced and an eight-faced figure, respectively.

Each of the six unit squares defined in the last paragraph corresponds to a four-period cycle of the elements in which the first and the last element alternately remains fixed. These squares are mutually disjoint and have a total of 24 vertices representing the 24 permutations of the four elements, see Figure 8.11.

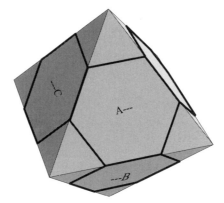

Figure 8.10   The eight hexagons joined to form a tetrakaidekahedron.

8.6   Four-Dimensional Models

Continuing with the geometrical analysis of the previous section, we find that the four-dimensional figure representing the permutation of five elements has ten tetrakaidekahedral faces corresponding to the $^5C_2 = 10$ end-constrained permutations of the five elements. Each of these ten tetrakaidekahedra has four hexagonal faces with fixed first and last elements and four hexagonal faces with fixed first and second elements or fixed fourth and fifth elements. The four hexagonal faces with fixed first and last elements match up with identical faces on other tetrakaidekahedra whilst the four hexagonal faces with fixed first two or last two elements pair up with similar hexagonal faces which have the same fixed elements in the same position but in reverse order. This second class comprises twenty hexagonal faces with their first two elements fixed and twenty with their last two elements fixed. In each case their twenty hexagonal faces define a set of ten plane tiles with hexagonal faces at unit distance from each other. Thus, we deduce that the four-dimensional polytope discussed in this paragraph has ten tetrakaidekahedra and two sets of ten hexagonal tiles as three-dimensional faces. Each of these tiles has twelve vertices and each set of ten tiles is mutually disjoint so that either set may be taken to represent the 120 permutations of five elements.

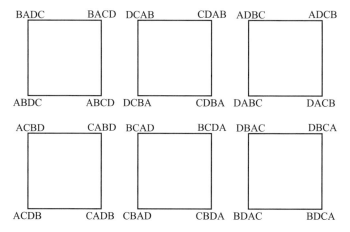

Figure 8.11 The six square faces of the tetrakaidekahedron.

It should be noted that this figure does not exist as a four-dimensional object with rigid edges. However, we are not restricted to conventional geometrical figures with rigid edges in this chapter and, in principle, we would have no difficulty constructing the figures representing the permutation of five or more elements from a system of flexible hollow tubes of unit length. This is left as an exercise for the reader!

8.7 Further Generalisations

In this chapter, we have shown that the mechanical models of Chapter 3 may be extended to two of the metric graph problems discussed by Monjardet (1991). These models are potentially valuable as they offer insights into the fundamental nature of the associated statistical problems. These models may readily be extended to the case where we are interested in choosing a point in the system of tubes which minimises the sum of the squared lengths of the paths by replacing the strings by springs. The point defined in this way is known as the centroid of the graph. Further variants of these models may be developed for the case in which we are interested in choosing one or more points in the system of tubes that jointly minimise the sum of the path lengths, or which jointly minimise the path length to the most distant node. The first of these problems is relevant when one is concerned with the location of social facilities, such as schools and libraries, where the access time may be averaged over all members of the community. By contrast, the second of these problems is relevant when one is concerned with the location of emergency facilities, such as hospitals and police, fire, and ambulance stations, where the maximum response time is the crucial factor.

References

Eilon, S., C. T. D. Watson-Gandy, and N. Christofides (1971), *Distribution Management: Mathematical Modelling and Practical Analysis*, Charles Griffin and Company, London.

Franksen, O. I. and I. Grattan-Guinness (1989), The earliest contribution to location theory? Spatio-economic equilibrium with Lamé and Clapeyron 1829, *Mathematics and Computers in Simulation* **31**: 195–220.

Hakimi, S. L. (1964), Optimum locations of switching centres and the absolute centres and medians of a graph, *Operations Research* **12**: 450–459.

Monjardet, B. (1991), Elements pour une histoire de la mediane metrique, in J. Feldman, G. Lagneau, and B. Matalon (Eds.), *Moyenne, Milieu, Centre: Histoires et Usages*, École des Hautes Études en Sciences Sociales, Paris, 45–62.

CHAPTER 9

Categorical Data Analysis

9.1 Nature of Categorical Data

As its name implies, categorical data analysis is concerned with the statistical analysis of sets of data which have been classified into one of several alternative categories for each of a number of qualitative characteristics. For simplicity, we shall restrict our discussion to the analysis of data sets which have been classified according to at most two characteristics. That is, we shall restrict ourselves to the analysis of one-way and two-way classifications. However, it should be noted that there is no such arbitrary restriction on the applicability of the mechanical models discussed in this chapter.

As a specific numerical example of the type of classification we have in mind, we follow Friendly (1995) by considering Snee's (1974) two-way classification of a set of 592 students by the colour of their eyes and the colour of their hair. The data on this group of students may conveniently be expressed in tabular form in Table 9.1.

The central portion of this table consists of a set of 16 cells arranged in the form of an array with four rows and four columns. This central portion is surrounded by a border consisting of a single row or column. The names in the leftmost column of the table indicate the eye colours to which the four rows of the central portion refer. Similarly, the names in the uppermost row indicate

Table 9.1   Two-way classification of 592 students

	HAIR COLOUR				
EYE COLOUR	Black	Brown	Red	Blond	TOTAL
Brown	68	119	26	7	220
Blue	20	84	17	94	215
Hazel	15	54	14	10	93
Green	5	29	14	16	64
TOTAL	108	286	71	127	592

the hair colours to which the entries in the four columns of the central portion refer. The rightmost column of the table sums the entries in the central portion over all hair colours and gives the total number of entries for each eye colour. Finally, the lowermost row sums the entries in the first four rows of the table and gives the total number of entries for each hair colour.

If the $c = 4$ columns of the table are added together, we obtain a one-way classification of the students by eye colour. Similarly, if the $r = 4$ rows of the table are added together then we obtain a one-way classification of the students by hair colour. For the present, we shall be concerned with the second of these two one-way classifications. The relevant data is expressed in tabular form as follows:

Table 9.2   One-way classification of 592 students

HAIR COLOUR				
Black	Brown	Red	Blond	TOTAL
108	286	71	127	592

This set of numbers may be visualised readily in the form of a bar chart or a pie diagram. A line of arbitrary length is cut into $c = 4$ portions whose lengths are proportional to the number of students in each class. If these line segments are placed side-by-side on a common baseline and at a common distance from each other, then we have the foundation of a standard bar chart or *histogram*, see Figure 9.1.

On the other hand, if they are differently shaded and placed end-to-end in a straight line then we have the basic form of a partitioned bar chart, see Figure 9.2.

Further, if the reconstructed straight line is bent round in a circle then we have a circular annulus which we may associate with the circumference of a pie diagram as indicated in Figure 9.3.

In the first two cases, the width of the line is irrelevant to the analysis and we may choose any suitable value. In principle, this statement also holds true for the line defining the circumference of a pie diagram; but, in practice, it is conventional to use this partitioned circular annulus as the basis for partitioning a solid disc into sectors of appropriate size, see Figure 9.4.

We shall not be concerned with the analysis of pie diagrams in this book as the physical analogies we require are more easily obtained from partitioned bar charts.

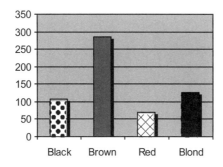

Figure 9.1 Side-by-side bar chart.

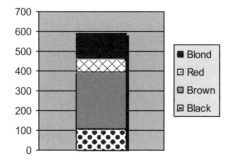

Figure 9.2 End-on-end bar chart.

9.2   Hydrostatic Models

9.2.1   Static Model

Although the standard and partitioned bar charts presented in Section 9.1 give a clear visual representation of the proportions of individuals in each of c classes, they are deficient for the purposes of statistical analysis as they are static and thus not readily associated with the concept of potential energy which we found so useful in earlier chapters. We shall consider the problem of side-by-side bar charts in this section and that of end-on-end bar charts in Section 9.3.

If we are given a set of n observations on a single characteristic which may fall into one of c categories, then this classification may

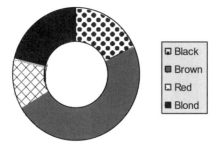

Figure 9.3 End-on-end bar chart in the form of a circular annulus.

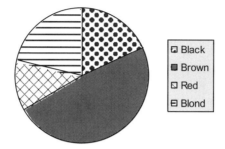

Figure 9.4   End-on-end bar chart in the form of a circular disc.

be represented by a set of c boxes or urns of uniform horizontal cross-section and indefinite height. We have to place n_1 observations in the first urn, n_2 observations in the second urn, and so on. We represent the n_i observations placed in the ith urn by n_i units of an incompressible substance. This incompressible substance may be solid, liquid, or gas; but it must flow easily and it must not mix with the surrounding gaseous medium. Thus, if the chosen substance is solid, then it must take the form of small particles such as ball-bearings, and, if it is a gas, then it must be heavier than the surrounding gaseous medium. In either case the chosen substance must behave as though it were an incompressible liquid and, for simplicity, we shall assume that it is actually such a liquid. We place n_i units of this liquid in the ith urn to produce a column of liquid whose height is proportional to the n_i observations in the ith class and thus, by spacing these columns of liquid evenly, we have a simple physical representation of a histogram, see Figure 9.5.

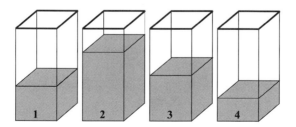

Figure 9.5   Urn model for the side-by-side bar chart.

As pointed out by Friendly (1995), this representation of the data by a histogram is entirely static. Even the chemical balance model of Chapter 1 has a dynamic component although it is not associated with an explicit optimality criterion.

9.2.2   Balance of Forces

For the next stage of our analysis, we assume that the urns or boxes have rectangular bases and that the liquid in the boxes may be adjusted to a common level by adjusting their width. For our purposes, we may assume that the heights of the boxes are arbitrarily large and that their depth is unity. For simplicity, we shall also assume that the total width of the boxes is unity, and that we may identify the right wall of each box with the left wall of its neighbour. In this way, we obtain a model which consists of a single box with unit width and unit depth divided by a set of $c - 1$ vertical partitions into c rectangular enclosures. The liquid in the separate enclosures may be adjusted to a common level by moving the interior divisions.

There will be a nonzero net force acting on an interior partition if the height of liquid on one side of the partition is different from that on the other. Thus, this model will be in a state of equilibrium when the level of liquid in each of the c boxes is the same. Now, there are a total of n units of liquid in the overall enclosure; and this overall enclosure has unit area, so that the height of liquid in each of the individual boxes must be n; and, the ith box must have width $p_i = n_i/n$ when this occurs.

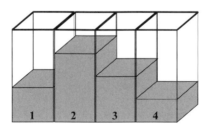

Figure 9.6   Hydrostatic model for the side-by-side bar chart.

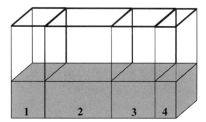

Figure 9.7 Hydrostatic model of the side-by-side bar chart when in equilibrium.

Clearly, the solution to the hydrostatic problem is obtained as a balance between the forces acting on the two sides of the $c - 1$ interior panels of the c individual boxes, as we shall now see.

9.2.3 Potential Energy Analysis

Consider a column of liquid of height h and horizontal cross-section a. Now, the uppermost slice of the column has area a and thickness dh at a height h above the base. Further, this slice has mass proportional to $2adh$, and thus potential energy proportional to $2ahdh$. Integrating this expression over all values of h from $h = 0$ to $h = h$, we find that the potential energy embodied in a column of liquid with base area a and height h is proportional to ah^2.

In our hydrostatic model, the ith column has area p_i and height n_i/p_i. Thus, the potential energy stored in the ith column of liquid is proportional to $p_i(n_i/p_i)^2 = n_i^2/p_i$, and we have to choose values for the parameters $p_1, p_2, \ldots, p_c$ in such a way as to minimise the potential energy in the system as a whole

$$P = \sum_i n_i^2/p_i$$

subject to the adding-up constraint on the overall widths of the boxes $\sum p_i = 1$.

Formulating the Lagrangian function

$$L = \sum n_i^2/p_i + \lambda\left(\sum p_i - 1\right)$$

and differentiating this function with respect to p_i, we find that we have to set $(n_i/p_i)^2$ equal to the (positive) constant $\lambda = \mu^2$. That is, we have to set p_i equal to n_i/μ, where

$$\mu = \mu \sum p_i = \sum n_i = n$$

The corresponding minimum value of the potential energy function is given by

$$P^* = n \sum_i n_i = n^2$$

thus the potential energy of the system exceeds this minimum value by the amount

$$P - P^* = \sum_i n_i^2/p_i - n^2$$

when it is in general position.

This difference may be written in a more convenient form. We consider the expression

$$\sum_i (n_i - np_i)^2/p_i = \sum_i (n_i^2/p_i - 2nn_ip_i/p_i + n^2p_i)$$

$$= \sum_i n_i^2/p_i - n^2$$

which reveals that our new hydrostatic model corresponds to a physical representation of the chi-squared optimality criterion

$$X^2 = \sum_i (n_i - np_i)^2/(np_i)$$

that is sometimes employed as an alternative to the likelihood criterion in categorical data analysis, see Berkson (1980).

9.2.4 Hypothesis Test of Independence

Setting $P_i = 1/c$ in the general expression of Subsection 9.2.3, we have

$$(P^{\#} - P^*)/n = \frac{c}{n}\sum_i n_i^2 - \sum_i n_i$$

for the amount of potential energy that is required when forcing the $c - 1$ movable panels to move from their equilibrium position with $p_i = n_i/n$, to the position in which they are equally spaced, $p_i = 1/c$. This expression is necessarily nonnegative and, when divided by n, may be used as the basis of a chi-squared test of the independence of the c classes, that is, of the hypothesis $H_0 : \pi_1 = \pi_2 = \ldots = \pi_c$ that the observations in each of the c classes occur with the same probability. To conduct this test, we have to compare the numerical value of this statistic with a suitable upper critical value of the chi-squared distribution with $c - 1$ degrees of freedom and reject the hypothesis under test if the numerical value is larger than the chosen critical value. In the case of the hair colour data of Section 9.1 this statistic takes the value $(P^{\#} - P^*)/n = 182.53$ and we may reject the hypothesis of independence at any conventional level of significance.

In this context, it is interesting to note that the $c - 1$ degrees of freedom of the chi-squared test are indicated by the $c - 1$ movable panels in the mechanical model.

9.3   Gas Pressure Models

9.3.1   Mechanical Model

Formalising the model of Section 9.2, we suppose that, for $i = 1$, $2, 3 \ldots, c$, we have n_i observations of an event with probability p_i, then the likelihood function associated with these $n = \sum_i n_i$ observations is given by the product

$$F = \prod (p_i)^{n_i}$$

and, taking logarithms in this expression, we have the log-likelihood function

$$\log(F) = \sum n_i \log(p_i)$$

which is clearly distinct from the optimality criterion developed in Subsection 9.2.3. Thus, the simple hydrostatic model of Section 9.2 is not a mechanical representation of the log-likelihood function.

To obtain a satisfactory analogue for this function, we have to replace the incompressible liquid or gas of Section 9.2 by a compressible gas or liquid. In this case, instead of replacing the n individual observations by unit volumes of liquid, we replace them by unit volumes of gas, such as those supplied in small cylinders of carbon dioxide used to fight automobile fires or to power soda syphons.

Suppose that the upper surface of the ith column of gas is secured in such a way that it contains n_i units of gas under unit pressure in a column of height n_i/p_i, width p_i, and depth unity. Now, suppose that we depress the upper boundary on this column in such a way that its height is reduced from n_i/p_i to unity. Then the pressure in the ith cell will increase from unity to n_i/p_i, and we have the basis of an alternative model for the statistical analysis of categorical data, see Figure 9.8.

We may derive the same result in a slightly different way: we place n_i volumes of gas in an enclosure with unit height and unit depth. If this cell also has unit width then we have n_i units of gas in a container of unit volume, so that the gas is subject to a pressure of n_i. If, further, the width of the ith cell is reduced from unity to p_i then we have the same volume of gas in a cell of volume p_i. Thus, the pressure of the gas on the walls of the ith cell is n_i/p_i per unit area.

Figure 9.8   Gas pressure model for the end-on-end bar chart.

In this context, we may note that a simple method for obtaining a representation of the pressure of the gas in the c cells is to suppose that each of the cylinders contains m molecules of gas, so that the ith cell will contain mn_i molecules of gas in volume p_i. Representing these mn_i molecules by points of a given size, we find that the ith cell contains mn_i points in an area of p_i units, so that the pressure of the gas per unit volume is represented in this figure by the density of the points per unit area, see Figure 9.9 and compare Figure 1.8 of Chapter 1.

9.3.2   Potential Energy Analysis

In Subsection 9.3.1, we established that the walls of the ith cell are subject to a pressure of n_i/p_i per unit area. The movable partition walls will clearly be in a state of equilibrium if the pressure on the two sides of each of these partitions is equal. Thus, we find that the ratio n_i/p_i must be constant, and hence that $p_i = n_i/n$ for $i = 1, 2, \ldots, c$.

As an alternative to this analysis based on the balance of forces in the model, we may integrate the expression n_i/p_i over all values of p_i from $p_i = p_i$ to $p_i = 1$ and find that the potential energy stored in the ith column of gas is $-n_i \log(p_i)$. Summing the c expressions of this type, we find that the potential energy in the system as a whole is given by

$$P = -\sum n_i \log(p_i)$$

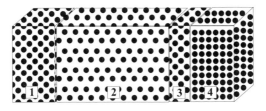

Figure 9.9   Relative density of gas molecules in two cells.

so that the potential energy function may be identified with the negative of the log-likelihood function of the corresponding statistical model

$$\log(F) = \sum n_i \log(p_i)$$

The optimal values of $p_1, p_2, \ldots, p_n$ are determined by choosing them to minimise this expression subject to the adding-up condition $\sum p_i = 1$. Formulating the Lagrangian function

$$L = - \sum n_i \log(p_i) + \lambda \left(\sum p_i - 1 \right)$$

and differentiating this expression with respect to p_i, we have $n_i/p_i = \lambda$ for all i whence $\lambda = \lambda \sum p_i = \sum n_i = n$ so that the optimal value of p_i is again given by $p_i = n_i/n$ for $i = 1, 2, \ldots, c$.

Substituting these values of p_i in the potential energy function

$$P = - \sum n_i \log(p_i)$$

we find that the minimum value of this function is given by

$$P^* = - \sum n_i \log(n_i/n)$$

and the difference between the two values by

$$P = P^* = - \sum n_i \log(n_i p_i/n)$$

9.3.3   Hypothesis Tests of Independence

If, instead of setting $p_i = n_i/n$, we set $p_i = 1/c$ as required by the hypothesis of independence of the c classes $H_0 \colon \pi_1 = \pi_2 = \ldots = \pi_c$ then we find that the difference in potential energy levels is given by

$$P^{\#} - P^* = \sum n_i [\log(c) + \log(n_i/n)] = - \sum n_i \log(n_i/(nc)),$$

and when multiplied by two, this expression defines the conventional likelihood ratio statistic for testing this hypothesis. To conduct this test, we again have to compare the numerical value of the statistic with a suitable upper critical value of the chi-squared distribution with $c - 1$ degrees of freedom and reject the hypothesis if the numerical value is larger than the chosen critical value. In the case of the hair colour data of Section 9.1, this statistic takes the value $2 = (P^\# - P^*) = 165.6$ and we may reject the hypothesis of independence at any conventional level of significance.

As in the case of the hydrostatic model of Section 9.2, we again note that the $c - 1$ degrees of freedom of the likelihood ratio test are indicated by the $c - 1$ movable panels in the gas pressure model. However, it is somewhat disconcerting to find, in the present context, that the conventional likelihood ratio statistic does not consist of a sum of squared terms. But this feature is easily explained in the context of the gas pressure model, as initially we have c cells of equal width $1/c$ with unequal amounts of gas in them whereas, when the system is in equilibrium, the size of the ith cell is proportional to the quantity of gas in that cell. Thus, to move from a state of equilibrium to the situation in which each of the cells has equal width, we have to increase the width of the narrower cells and reduce that of the wider cells. The widening of the narrower cells will reduce the pressure in these cells and thus result in a negative contribution to the potential energy function. By contrast, the reduction in the width of the wider cells will result in an increase in pressure and a positive contribution to the potential energy function. It may not seem intuitively obvious, but the structure of the physical model assures us that the sum of the positive contributions to the likelihood ratio statistic must exceed that of the negative contributions, as the equilibrium state represents a minimum value for this function.

9.3.4 Analogy with the Chemical Balance Model

In passing, we note that the hydrostatic and gas pressure models developed in this chapter may be interpreted as generalisations of the simple chemical balance model of Chapter 1 with a weight of n_1 units on one side of the fulcrum and a weight of n_2 units on the other side. We suppose that the weights are located at the points $y = y_1$ and $y = y_2$ and that the fulcrum is placed at the point $y = a$ where $y_1 < a < y_2$. We set $p_1 = (a - y_i)/(y_2 - y_1)$ and $p_2 = (y_2 - a)/(y_2 - y_1)$. In this notation, the balance of rotational forces about the fulcrum is given by the condition $n_1 p_1 = n_2 p_2$. This condition is to be contrasted with the corresponding conditions for the balance of forces in the hydrostatic and gas pressure models indicated by the equations $n_1/p_1 = n_2/p_2$ and $-n_1 \log(p_1) = -n_2 \log(p_2)$ respectively. In particular, it should be noted that the chemical balance model of Chapter 1 uses an increasing function of p_i (actually the identity function) whereas the models of the present chapter use decreasing functions. This difference in structure is indicated in the different forms of their solution: the larger the value of the ratio n_1/n_2, the smaller is the value of p_1 in the chemical balance model and the larger is the value of this parameter in the hydrostatic and gas pressure models.

9.4   Two-Way Tables

The hydrostatic and gas pressure models of earlier sections may readily be extended to higher dimensions. In this section, we shall restrict our attention to the two-dimensional case. In this context, we have to place n_{ij} volumes of liquid or gas in the ijth cell of a two-dimensional array. In the case of the hydrostatic model the potential energy function is given by

$$P = \sum_i \sum_j n_{ij}^2/p_{ij}$$

and in the case of the gas pressure model it is given by

$$P = -\sum_i \sum_j n_{ij} \log(p_{ij})$$

In either case, these expressions are minimised subject to the adding-up constraint $\sum_i \sum_j p_{ij} = 1$ by setting p_{ij} equal to n_{ij}/n where $n = \sum_i \sum_j n_{ij}$. These are the unconstrained values.

   If the column partitions are removed or the cells in the rows are connected by tubes, then the level of the liquid or the pressure of the gas will adjust to a common level which in turn indicates the total volume of liquid or gas in each row. In this context, we have r rows whose potential energy is minimised by setting $p_{i*} = n_{i*}/n$ where $n_{i*} = \sum_j n_{ij}$.

Similarly, if the partitions in the rows of the model are removed or the cells of the individual columns are otherwise connected so that the liquid or gas in each of the columns is able to adjust to a common level, then the levels in the c columns will correspond to the total numbers of individuals recorded in each column of the table. Thus we have a set of c columns whose potential energy is minimised by setting p_{*j} equal to n_{*j}/n where $n_{*j} = \sum_i n_{ij}$.

If the row and column effects are independent, then we may hypothesise that $\pi_{ij} = \pi_{i*}\pi_{*j}$ for all values of i and j. The potential energy function is minimised subject to this condition by setting p_{ij} equal to the product of the estimates, that is, by setting p_{ij} equal to $n_{i*}n_{*j}/n^2$.

Substituting these values in the hydrostatic potential energy function, we have

$$P^{\#} = \sum_i \sum_j n_{ij}^2 n^2/(n - i*n*j)$$

so that the gain in potential energy in moving from the unconstrained model to this independent effects model is given by the expression

$$P^{\#} - P^{*} = n^2 \left[\sum \sum n_{ij}^2/(n_{i*}n_{*j}) - 1 \right]$$

Similarly, for the gas pressure model we find that the potential energy function takes the value

$$P^{\#} = - \sum_i \sum_j n_{ij} \log(n_{i*}n_{*j}/n^2)$$

and thus the difference in potential energy levels in moving from the unconstrained values to those associated with the independent effects model is given by

$$P^{\#} - P^{*} = - \sum_i \sum_j n_{ij} \log[(n_{i*}n_{*j})/(nn_{ij})]$$

This expression corresponds to the likelihood ratio test statistic described by Friendly (1995) who uses it to conduct a series of tests of the hair and eye colour data of Section 9.1.

In the circumstances envisages in this chapter, it is possible to obtain suitable estimates of the probabilities π_i or π_{ij} without difficulty. In higher dimensions, and in the context of more intricate models, iterative solution procedures will have to be employed. The hydrostatic and gas pressure models offer insights into the basis of suitable iterative solution procedures for such problems, see Sall (1991) or Friendly (1995) for details.

9.5   Hydrostatic Models in Economics

A hydrostatic model which is closely related to the model of Section 9.2 was developed by Irving Fisher in 1892 to illustrate the mechanisms of the neoclassical model of economic equilibrium. In this model, the prices of the economic system are normalised by setting one of them (known as the *numéraire*), rather than their sum, equal to unity. Further, the model imposes a system of marginal utility constraints by means of floats which are connected to one another by mechanical arms. The first of these amendments is of no real significance but the second represents a major modification of the model presented in Section 9.2.

References

Berkson, J. (1980), Minimum chi-square, not maximum likelihood, *Annals of Statistics* **8**: 457–469.

Fisher, I. (1892), *Mathematical Investigations in the Theory of Value and Price*, Yale University Press, New Haven, Connecticut. Reprinted by Augustus M. Kelly, New York, 1961.

Friendly, M. (1995), Conceptual and visual models for categorical data, *The American Statistician* **49**: 153–160.

Sall, J. (1991), The conceptual model for categorical responses, *ASA Statistical Computing and Statistical Graphics Newsletter* **3**: 33–36.

Snee, R. D. (1974), Graphical display of two-way contingency tables, *The American Statistician* **28**: 9–12.

CHAPTER 10

Method of Averages and Curve Fitting by Splines

10.1 Mechanical Models for Multivariate Means

10.1.1 One-Dimensional Means

In Section 1.4, we developed a simple mechanical model for the arithmetic mean of a set of observations based on a chemical balance in a state of equilibrium. Given a set of n direct observations on a single variable Y, we represented these n observations as a scatter of points on the y-axis in the horizontal plane and treated this line as if it were a horizontal beam with unit weights suspended from the given points. Our problem was to locate the point of balance of the weighted beam. Arbitrarily selecting a point on the line and placing a fulcrum at this point, we found that weights to the right of this point will cause the beam to rotate in a clockwise direction about the fulcrum whilst weights to the left will cause it to rotate in an anticlockwise direction. We have thus to multiply the value of the ith weight by the distance from the fulcrum to the point of suspension to determine the ith weight's contribution to the net clockwise couple. Summing these contributions, we may determine the net clockwise couple of the system as a whole about the fulcrum. If this value is positive then the beam will rotate in a clockwise direction, and if negative it will rotate in an anticlockwise direction, see Figure 10.1a. Zero values of this function will identify the point of balance of the system of unit weights. This point of balance is usually known as the centre of gravity or the centroid of the system and is located at the Arithmetic Mean of the n given observations, see Figure 10.1b.

10.1.2   Multidimensional Means

The technique outlined in Subsection 10.1.1 may readily be extended to higher dimensions. However, for simplicity, we shall restrict our discussion to the two-dimensional case. Suppose that we are given a set of n points scattered in the two-dimensional horizontal plane and that our problem is to identify the centre of gravity of these n points. We specify an arbitrary direction in the plane, and consider the effect of couples taken about an arbitrary line in this direction. Couples are calculated in the same way as in Subsection 10.1.1. We multiply the weight of the ith observation by the perpendicular distance from the ith point to the arbitrary line, and separately sum the positive and negative couples about this line. If these two partial sums do not balance then they may be brought into closer agreement by moving the arbitrary line parallel to itself in the direction of the larger partial sum.

    In dimensions greater than two, the line of balance in a given direction is replaced by a $(q - 1)$-dimensional hyperplane of balance in a given direction. The direction of the arbitrary hyperplane is most easily specified by identifying a line which is at right angles to all lines in this hyperplane. The distances from each of the n given points to the arbitrary hyperplane are then marked off on this line. And the optimal position for the hyperplane in the spe-

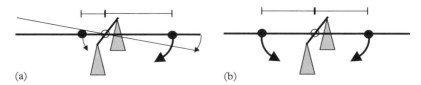

(a)   (b)

Figure 10.1   A loaded beam (a) when not in equilibrium, and (b) when in equilibrium.

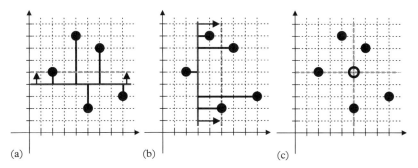

Figure 10.2 (a) The line of balance parallel to the y-axis, (b) the line of balance parallel to the x-axis, and (c) their point of intersection.

cified direction is found to correspond to the mean of the projected points. Further, the point of balance for the system as a whole is readily obtained by considering a set of q such families of hyperplanes.

This alternative approach to the fitting problem is illustrated in Figures 10.3a and 10.3b for the two-dimensional case in which the arbitrary one-dimensional hyperplane is drawn parallel to the x-axis. An arbitrary line is drawn at right angles to this hyperplane and parallel to the y-axis; the distances between the observed points and the arbitrary hyperplane are then marked off on this line and their mean determined, see Figure 10.3a. The optimal hyperplane is then obtained by moving the arbitrary hyperplane parallel to itself in such a way as to pass through this mean value as indicated in Figure 10.3b.

10.2 The Method of Averages

In Subsection 10.1.2, we have established that a horizontal plane loaded with a system of unit weights at n given points will balance at the centre of gravity or centroid of this system of weights. A second point of balance may be obtained by employing a second system of weights. Further, if the two points determined in this way are joined by a straight line then we obtain a line of fit to the n given points. In general,

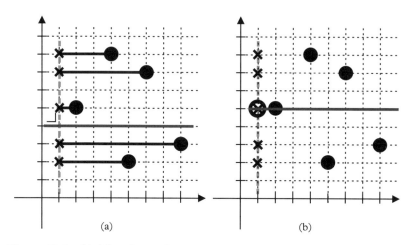

Figure 10.3 (a) The observed points projected onto an arbitrary line in the
y-direction, and (b) the resulting line of balance in the x-direction.

this method of fitting a line to a set of points is known as the
method of averages. It is also known as the method of instrumental variables when the q sets of weights employed in the
analysis are determined by a second system of explanatory
variables called instrumental variables.

In this chapter, we shall restrict our discussion to the original
form of the method of averages. In this case, the observations are
divided into two or three groups of roughly equal size, the means
of the two extreme groups are determined, and a straight line is
drawn through these two means.

Consider the $n = 15$ observations in Table 10.1. These observations may be divided into three groups of equal size by those

Table 10.1
Artificial data based on the population of the U.S.A.

X	1	2	3	4	5	6	7	8	9	10	11	12	13	14	15
Y	128	170	231	314	397	501	629	759	919	1057	1227	1318	1506	1793	2031

with X-values below $X = 5.5$, those with X-values between $X = 5.5$ and $X = 10.5$, and those with X-values above $X = 10.5$. The means of the observations in the first group are $\bar{x} = 3$ and $\bar{y} = 245$ whilst the corresponding values for the third group are $\bar{x} = 13$ and $\bar{y} = 1575$. The point corresponding to these centres of gravity are marked in Figure 10.4 together with the line joining them.

Some authors are willing to employ the line determined in this way as the line of best fit to the data. Others prefer to plot the centre of gravity $(\bar{x}, \bar{y}) = (8, \ 865.3)$ of all $n = 15$ observations and to draw a line through this point parallel to the original line. That is, they are willing to use the slope of the original line but prefer to displace it to pass through the overall centre of gravity, see Figure 10.5, noting that the fitted line passes through the point $(x, y) = (8, \ 910)$ in Figure 10.4.

Either variant of the original method of averages offers a simple procedure for determining preliminary estimates of the parameters of a fitted relationship, see Farebrother (1998).

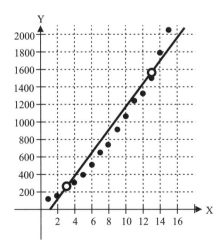

Figure 10.4 The line joining the centroids of the extreme groups.

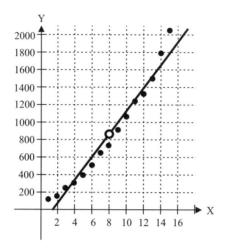

Figure 10.5   The line of Figure 10.4 displaced to pass through the overall centroid.

10.3   Smoothing by Linear and Cubic Splines

10.3.1   Linear Splines

If, instead of determining a line of best fit by passing a single line through the centroids of the two extreme groups, we evaluate the centroids of all three groups and pass a line segment through each of these centroids, then we have the basis of an alternative fitting procedure. In particular, we may choose these line segments in such a way as to minimise the sum of the squared deviations within each of the three groups as illustrated in Figure 10.6.

An interesting variant of this procedure is obtained if we minimise the sum of the squared deviations over all three groups subject to the requirement that the right end of each line segment should meet the left end of the next segment on a fixed line defining the boundary between the two groups. Thus, in Figure 10.7 we have five observations satisfying the relationship

$$y_i = a_1 + b_1 x_i + e_i \quad i = 1, 2, \ldots, 5$$

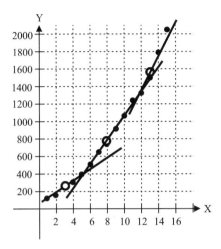

Figure 10.6 Distinct line segments fitted to each of the three groups.

five observations satisfying

$$y_i = a_2 + b_2 x_i + e_i \quad i = 6, 7, \ldots, 10$$

and five observations satisfying

$$y_i = a_3 + b_3 x_i + e_i \quad i = 11, 12, \ldots, 15$$

where a_1, a_2, a_3 are the intercept parameters and b_1, b_2, b_3 are the slope parameters in the three groups. In this context, our problem is to choose values for these six parameters in such a way as to minimise the sum of the squared deviations $\sum e_i^2$ subject to the linear constraints

$$a_1 + b_1 k_1 = a_2 + b_2 k_1$$

and

$$a_2 + b_2 k_2 = a_3 + b_3 k_2$$

where $X = k_1$ defines the line dividing the first group of observations from the second, and $X = k_2$ defines the line dividing the second group from the third. If the values of the $m - 1$ *knots* k_1 and k_2 are known, as we shall suppose, then this method of curve fitting by *linear splines* takes the form of a standard least squares fitting problem subject to a set of $m - 1$ linear constraints of the type discussed in Section 4.6. Although the joint fit of the individual line segments in Figure 10.7 may well be better than those of the lines illustrated in Figures 10.4 and 10.5, the presence of these linear constraints means that the fit illustrated in Figure 10.7 is necessarily not as close as that illustrated in Figure 10.6.

A mechanical model for this problem may easily be developed from the models of Chapters 3 and 4. A set of $m - 1$ rigid rods are fixed in a south–north direction on a horizontal board in positions corresponding to the values of the $m - 1$ given knots. A small ring is placed over each of these fixed rods. Then a rigid rod is passed through the ring on the first fixed rod, a second rigid rod is passed through the rings on the first and second fixed rods, a third rigid rod is passed through the rings on the second and third fixed rods, ..., and an mth rigid rod is passed through the ring on the last fixed rod. Finally, the observations in each of the m groups are connected to the appropriate movable rod by springs of unit modulus and zero natural length running parallel to the set of fixed rods. As in Chapter 4, the fit determined by the method of linear splines corresponds to the minimum potential energy level of this system of rods and springs.

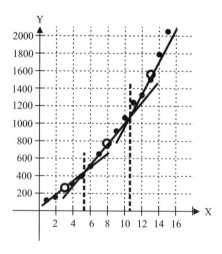

Figure 10.7 Spline fit of linear functions to the three groups.

10.3.2 Cubic Splines

The method of curve fitting described in Subsection 10.3.1 suffers from the obvious defect that the joins between the successive groups of observations are rather abrupt. However, this problem is easily surmounted by replacing the linear relationship

$$y_i = a_j + b_j x_i + e_i$$

describing the positions of observations in the jth group by the corresponding cubic equation

$$y_i = a_j + b_j x_i + c_j x_i^2 + d_j x_i^3 + e_i$$

The resulting *cubic spline* fitting procedure is obtained if we minimise the sum of the squared deviations over all m groups subject to the requirement that the right end of each cubic segment should meet the left end of the next segment in a sufficiently smooth manner on the fixed line defining the boundary between these two groups. That is, we have to minimise the sum of the squared deviations $\sum e_i^2$ subject to a set of $3(m-1)$ constraints of the form

$$a_1 + b_1 k_1 + c_1 k_1^2 + d_1 k_1^3 = a_2 + b_2 k_1 + c_2 k_1^2 + d_2 k_1^3$$
$$b_1 + 2c_1 k_1 + 3d_1 k_1^2 = b_2 + 2c_2 k_1 + 3d_2 k_1^2$$
$$2c_1 + 6d_1 k_1 = 2c_2 + 6d_2 k_1$$

where $X = k_1$ again defines the line dividing the first group of observations from the second. Note that the second of these constraints is the derivative of the first, and that the third is the derivative of the second. Once again, if the values of the knots $k_1, k_2, \ldots, k_{m-1}$ are known, then this method of curve fitting by cubic splines takes the form of a least squares curve fitting problem subject to a set of $3(m-1)$ linear constraints, see Poirier (1976) or Pollock (1999).

Figure 10.8 indicates the result of fitting a curve to the data in Table 10.1 by the method of cubic splines. Comparing these results with those illustrated in Figure 10.7, it will be apparent that the method of cubic splines is better able to identify an underlying

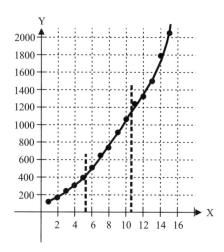

Figure 10.8   Spline fit of cubic functions to the three groups.

curvilinear relationship between two variables than is the method of linear splines. Indeed, in Figure 10.8 we have an almost perfect fit to the data.

A mechanical model for this problem may easily be developed from the model used in Subsection 10.3.1. A set of $m-1$ rigid rods are again fixed in a south–north direction on a horizontal board. A ring is placed over each of these fixed rods, and a single *flexible* rod is passed through the $m-1$ rings on the $m-1$ fixed rods. Finally, the n observations are connected to the single flexible rod by springs running parallel to the fixed rods. Once again, the optimal position of the flexible rod is determined by the minimal value of the potential energy function. However, a formal potential energy analysis of this model is complicated by the quantity of potential energy stored by the bending of the flexible rod, and therefore we shall not examine this problem in any greater detail in this book.

In passing, it is worth noting that the flexible rod employed in this model is responsible for the name of the fitting procedure itself, as the flexible plastic or wooden rod used by engineers when preparing their technical drawings are known as *splines*.

10.4 Multidimensional Medians

In this chapter, we have restricted our discussion to optimality criteria based on the sum of the squared deviations criterion. These models are easily extended to other optimality criteria. For example, as a simple generalisation of the method of averages discussed in Section 10.2, we may replace the centroids of the groups of observations by their mediancentres defined in Section 3.2. Thus, we may define a simple variant of the method of averages by dividing the data set into three groups of observations, computing the mediancentres of the two extreme groups, and joining these mediancentres by a straight line. However, it is relatively difficult to determine the mediancentre of a large set of points, and we may prefer to substitute the componentwise median for the mediancentre in the formulation of this procedure. That is, we may prefer to use the point determined by combining the median of the x-coordinates with the median of the y-coordinates in each group.

In the case of the basic fitting procedure, the line through the componentwise medians of the two groups of points is taken as the line of best fit to this set of data. However, some authors prefer to replace this line by the line parallel to it which passes through the componentwise median of the data set as a whole. This second line of fit is known as Tukey's resistant line, see Emerson and Hoaglin (1983) or Johnstone and Velleman (1985) for a detailed discussion of this robust fitting technique.

In general, the componentwise medians employed in this procedure will not coincide with observations in the data set. But, because of the special nature of the data in Table 10.1, we find that the componentwise medians for the first, third, and complete data sets correspond to the third, thirteenth, and eighth observations respectively. The alternative lines of best fit defined by the procedures of the previous paragraph are illustrated in Figure 10.9a and 10.9b.

The results obtained here should be compared with those illustrated in Figures 10.4 and 10.5.

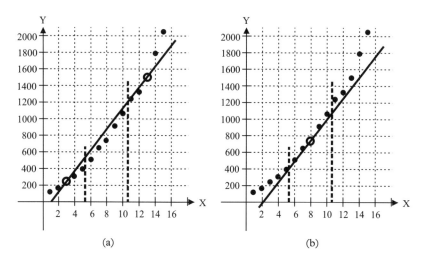

Figure 10.9 (a) Line joining the pairwise medians of the extreme groups, and (b) the line displaced to pass through the overall pairwise median.

References

Emerson, J. D. and D. C. Hoaglin (1983), Resistant lines for y versus x, in D. C. Hoaglin, F. C. Mosteller and J. N. Tukey (eds.), *Understanding Robust Exploratory Data Analysis*, John Wiley and Sons, New York, 129–165.

Farebrother, R. W. (1998), Nonlinear curve fitting and the true method of least squares, *The Statistician* **47**: 137–147.

Johnstone, I. M. and P. F. Velleman (1985), The Resistant Line and Related Regression Methods, *Journal of the American Statistical Association* **80**: 1041–1054.

Poirier, D. (1976), *The Econometrics of Structural Change with Special Emphasis on Spline Functions*, North-Holland Publishing Company, Amsterdam.

Pollock, D. S. G. (1999), *Time-Series Analysis, Signal Processing and Dynamics*, Academic Press, New York.

CHAPTER 11

Multivariate Generalisations of the Method of Least Squares

11.1 Multivariate Statistical Analysis

Thus far in this book, we have been largely concerned with the problem of fitting a set of observations on a single dependent variable to a set of observations on one or more explanatory variables. The principal exceptions to this rule occur in the body of Chapter 3, and in the introductory sections of Chapters 4 to 7, where we were concerned with the reverse problem of fitting a set of observations on two or more jointly dependent variables to a set of unit observations on a single constant explanatory variable. In this chapter, we shall examine the possibility of extending our mechanical ideas to other areas of multivariate statistical analysis. Of course, we shall not be able to develop the selected multivariate techniques in full detail and we therefore refer interested readers to one of the many excellent textbooks on the subject.

11.2 Orthogonal Least Squares and Principal Components

11.2.1 Orthogonal Least Squares

Suppose that we are given a set of n observations on each of a set of q variables. Then we may either represent these n observations as a set of q points in n-dimensional prediction space, or we may represent them as a set of n points in q-dimensional observation space. We briefly discussed the first of these approaches to the problem in Chapter 5; here we shall address the second option.

Representing each of the n observations on the q variables as a single point in q-dimensional space, we obtain a set of n points in this space. For simplicity, we assume that we have observations on $q = 2$ variables so that these n observations may be plotted in a two-dimensional plane. Our first problem is that of determining a single point which best represents the position of these n points. We therefore choose an arbitrary point in the plane and adjust its position in such a way as to minimise the sum of the squared distances from the n given points to the additional point. As we have seen in Chapter 4, this criterion is minimised when the arbitrary point is placed at the centre of gravity or centroid of the swarm of points, see Figure 11.1.

Our next problem is that of choosing a line which best represents the long axis of the swarm of points. Placing an arbitrary straight line in the horizontal plane and joining each of the points in this plane to the arbitrary line by a line segment which lies perpendicular to the arbitrary line, we have to choose the position of the fitted line in such a way as to minimise the

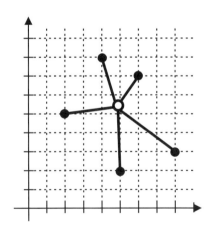

Figure 11.1   Least squares fit of a point.

sum of the squared perpendicular distances from the n given points to this line, see Figure 11.2. This problem defines the orthogonal least squares fit of the line to the data as outlined in Section 4.3.

This orthogonal line fitting problem may readily be generalised to higher dimensions. In the case when q is greater than two, we may be interested in fitting a two-dimensional plane, a three-dimensional hyperplane, ..., or a $(q - 1)$-dimensional hyperplane to the data, and to do this in such a way that the sum of the squares of the perpendicular distances from the n given points to the fitted plane or hyperplane is minimised. Figure 11.3 illustrates the particular case when we are interested in fitting a two-dimensional plane to a set of n points in three-dimensional space.

It is a well known feature of the orthogonal least squares problem that the optimal point lies on the optimal line, the optimal line lies on the optimal plane, and so on. It is therefore possible to construct these orthogonal least squares fits in a sequential manner. First we find the optimal point; then we find the optimal line passing through this point; then we find a second line through the optimal point, which when combined with the first line defines the optimal plane; then we find a third line through the optimal point, which when combined with the first two lines defines the optimal three-dimensional hyperplane, and so on. If the successive lines of

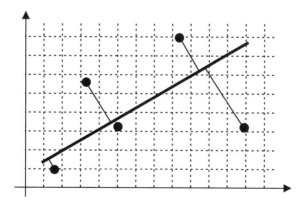

Figure 11.2   Orthogonal least squares fit of a line to a set of points.

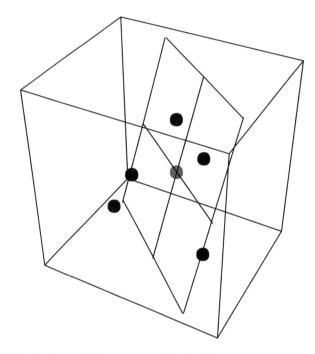

Figure 11.3 Orthogonal least squares fit of a plane.

best fit are at right angles to each other, then these lines are associated with the successive principal axes of the swarm of points;
they are therefore known as the *principal components* of the data
set. (In practice, this set of principal components and their associated sums of squared deviations are usually determined from the
eigenvectors and eigenvalues of the matrix of squares and cross-
products of the given observations expressed in deviations from
their sample means. However, it will be clear from the discussion in
Subsection 4.3.3 that a detailed examination of this approach to
the fitting problem would take us well beyond the intended scope
of this book. Interested readers are therefore referred to the relevant section of their preferred textbook on multivariate statistics
for details.)

     In passing, we note that the sequential technique outlined in
the previous paragraph is often employed when the fitting criterion
is *not* that of orthogonal least squares. This misapplication of the

rule naturally leads to a disassociation between the nominal fitting criterion and the one actually being used.

Finally, we note that in Chapter 4, we discussed a mechanical model for the orthogonal least squares fitting of a straight line to a set of points in the horizontal xy-plane. We drew an arbitrary line in this plane and attached a set of springs running perpendicularly from the given points to a rod lying on the chosen line. When the rod is in a state of equilibrium, it minimises the potential energy in the stretched springs and thus identifies the position of the orthogonal least squares line.

11.2.2   Fitting Elliptical Contours

An alternative realisation of the principal components problem is obtained by supposing that the observations on the variables X and Y come from an elliptically symmetric distribution, such as the bivariate normal distribution. In this context, our problem is to determine a system of concentric elliptical contours which seem best to fit the observed data. Having selected a system of concentric ellipses, we may readily determine the lines of symmetry and the two conventional lines of tangents as illustrated in Figures 1.5, 1.6, and 1.7 of Chapter 1. Clearly, there is most variation about the minor axis of the typical ellipse, and least about its major axis, which thus define the larger and the smaller principal components respectively. In addition, the major axis of the ellipse corresponds to the orthogonal least squares line mentioned above.

As a first illustration of this problem, we consider the data given in Table 2.1 and plotted in Figure 2.10 of Chapter 2. The orthogonal least squares line, and therefore the major axis of the system of ellipses, runs in a northeasterly direction, but it is difficult to see how we can make further progress in the analysis of this data set without using the more advanced techniques of matrix algebra. The nature of this fitting problem becomes somewhat clearer when the observations on the variables X and Y are cross-classified into a number of grid squares in the xy-plane. For our second example, we consider the data given in Table

11.1, in which the heights (in inches) of 928 adult children and the average of the heights of their 205 mothers and 205 fathers have been cross-classified. (In this context, all female heights have been multiplied by a factor of 1.08 to make them comparable with the corresponding male heights.)

For the purpose of this example, the relevant section of Table 11.1 is the portion that remains after the first and last rows and the first and last columns have been deleted. An examination of this portion of the table reveals that its contours are elliptical in form with a common major axis running in a northwesterly direction. We shall leave readers to consider how they might best fit a system of concentric ellipses to this data set without employing the advanced techniques of modern mathematical statistics. For descriptions of Francis Galton's 1885 solution to this problem, and the associated discovery of regression lines and correlation coefficients, see Hald (1998, pp. 608–616) or Stigler (1986, pp. 283–290).

11.2.3 Fitting a Circle

As a simple generalisation of the problem of fitting a straight line to a set of n points, we consider the problem of fitting a circle to a set of n points in the horizontal xy-plane. We suppose that we have a scatter of n points in the xy-plane. We select an arbitrary point in this plane, draw a circle with arbitrary radius about the chosen point, and attach springs running perpendicularly from the n given points to the circle. In this context, we have to determine the net forces pulling in the south–north and west–east directions for a fixed circle, and the net forces pulling radially for a fixed centre. Clearly, zero values for all three functions will determine the optimal orthogonal least sum of squared deviations circle, see Figures 11.4 and 11.5.

This mechanical model and the corresponding model for the fitting of a point (or a circle of zero radius) may readily be generalised to spherical data sets as discussed in Subsection 11.3.2 below.

Table 11.1 Cross-classification of the heights of 928 adult children and the heights of their 205 midparents

Height of midparent	Height of adult child														Total
	<61.7	62.2	63.2	64.2	65.2	66.2	67.2	68.2	69.2	70.2	71.2	72.2	73.2	>73.7	
>73.0	–	–	–	–	–	–	–	–	–	–	–	1	3	–	4
72.5	–	–	–	–	–	–	–	1	2	1	2	7	2	4	19
71.5	–	–	–	–	1	3	4	3	5	10	4	9	2	2	43
70.5	1	–	1	–	1	1	3	12	18	14	7	4	3	3	68
69.5	–	–	1	16	4	17	27	20	33	25	20	11	4	5	183
68.5	1	–	7	11	16	25	31	34	48	21	18	4	3	–	219
67.5	–	3	5	14	15	36	38	28	38	19	11	4	–	–	211
66.5	–	3	3	5	2	17	17	14	13	4	–	–	–	–	78
65.5	1	–	9	5	7	11	11	7	7	5	2	1	–	–	66
64.5	1	1	4	4	1	5	5	–	2	–	–	–	–	–	23
<64.0	1	–	2	4	1	2	2	1	1	–	–	–	–	–	14
Total	5	7	32	59	48	117	138	120	167	99	64	41	17	14	928

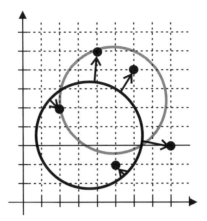

Figure 11.4 Least squares fit of a circle – initial position.

11.3 Procrustes Rotation

11.3.1 Orthogonal Procrustes Rotation

Procrustes rotations of various types are named for the giant in Greek mythology who adjusted the length of his victims to fit the size of his bed. In this section, we shall discuss a technique known

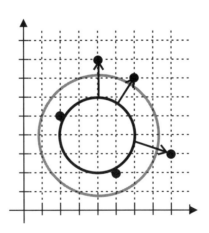

Figure 11.5 Least squares fit of a circle – final position.

as orthogonal Procrustes rotation. Suppose that we have two sets of n observations on the same q variables. Each of these sets of observations may be represented as sets of n points in q-dimensional space. Having determined the centroid of each set and adjusted both sets of observations so that these centroids are placed at the origin, our problem is that of rotating the second set of observations about its centroid in such a way that it most nearly approaches the position of the first set. If the jth element of the ith observation in the first set is denoted by y_{ij} and the corresponding element in the set of observations after it has been rotated through an arbitrary angle is denoted by z_{ij}, then a common measure of distance is the square root of the sum of the squared deviations between these two quantities

$$\sum \sum (y_{ij} - z_{ij})^2$$

We note that the ijth component in the first set is paired with the corresponding component in the transformed set. We may therefore readily establish a simple mechanical model for this procedure by plotting the two sets of data on two planes with the same axes and the same scales. Joining these two planes at the points corresponding to the centroids of the two data sets, we attach one end of an ideal spring of unit modulus to the ith point on the plane representing the first set of observations and the other to the ith point on the plane representing the second set.

As illustrated in Figures 11.6 and 11.7, our problem becomes one of rotating the second plane relative to the first in such a way as to minimise the potential energy in the system of stretched springs; that is, the sum of the squared distances between the two systems of points.

11.3.2   Fitting Spherical Data

If each of the observations in each of the two sets of data in Subsection 11.3.1 lies at unit distance from the corresponding centroid, then the orthogonal Procrustes problem becomes one of fitting a set of n observations on the surface of a unit sphere to a second set of n observations on the surface of a second unit sphere.

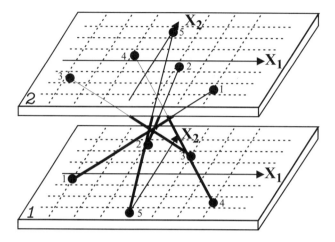

Figure 11.6   Orthogonal Procrustes rotation – initial position.

This is the fitting problem with directional data discussed by Jupp and Mardia (1980).

The problem of fitting points and circles to sets of observations located on the surface of a unit sphere arises naturally when one is interested in the relative position of geographical features on the surface of the Earth, the Moon, or one of the planets, or in the

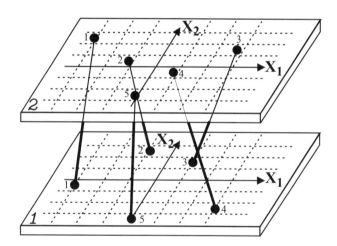

Figure 11.7   Orthogonal Procrustes rotation – final position.

relative position of astronomical objects against the background of the celestial sphere. In this context, it should be mentioned that the methods employed in Chapters 4 and 5 to fit a point or a straight line to a set of observations in a plane may readily be generalised to the fitting of a point or a small circle to a set of observations on the surface of a sphere, see Fisher (1985) and Mardia and Gadsden (1977). The only difference is that the lengths of the strings or springs making up the mechanical models must lie on the surface of a sphere rather than on the surface of a horizontal plane, see Farebrother (1992).

11.4   Multidimensional Scaling

The multidimensional scaling problem is similar to the orthogonal Procrustes problem discussed in Section 11.3 except that we no longer have direct access to sets of observations on two variables in q-dimensional space. Instead, we have a complete set of measurements of the distances between every pair of points in a single set of observations in q-dimensional space, and our problem is to determine whether these measurements can be closely approximated by the corresponding distances between pairs of points in a Euclidean space of suitable dimension. Table 11.2 gives the road distances in miles between a selection of nine British university towns, where Abd = Aberdeen, Aby = Aberystwyth, Edi = Edinburgh, Exe = Exeter, Gla = Glasgow, Lon = London, Man = Manchester, Nor = Norwich, and Oxf = Oxford. The distances given in this table are road distances rather than great circle distances, so it is not clear they can be closely approximated by the straight line distances between the corresponding pairs of points on a plane map. In this context, it would be usual to illustrate the problem by supplying an outline map of the mainland of Britain. However, we have not done so in order that readers may attempt to reconstruct the relative positions of these nine towns given that Edinburgh is to the north of London, and that Aberystwyth is to the west of the Great North Road joining Edinburgh and London.

   To construct a mechanical model for this problem, we represent the positions of the n towns by n points on a horizontal plane

**Table 11.2 Road distances between nine
British university towns**

	Abd	Aby	Edi	Exe	Gla	Lon	Man	Nor	Oxf
Abd	0	470	125	589	148	548	355	488	505
Aby	470	0	335	206	334	238	131	277	159
Edi	125	335	0	454	46	413	220	360	370
Exe	589	206	454	0	454	200	246	313	154
Gla	148	334	46	454	0	411	218	380	368
Lon	548	238	413	200	411	0	203	115	56
Man	355	131	220	246	218	203	0	185	160
Nor	488	277	360	313	380	115	185	0	143
Oxf	505	159	370	154	368	56	160	143	0

and compute the distance between the ith and the jth of these points. Denoting this distance by δ_{ij} and the corresponding measured physical distance by d_{ij}, it seems reasonable to choose the second system of points to minimise the sum of the squared deviations between the two sets of measurements, that is, to minimise the expression

$$\sum\sum(\delta_{ij} - d_{ij})^2$$

If we correct the ith point in the graphical scheme to the jth point by a spring of natural length d_{ij} and unit modulus, then we find that it contributes the correct amount of potential energy to the overall potential energy function provided that the spring is under tension, that is, provided that δ_{ij} is greater than d_{ij}. If, on the other hand, δ_{ij} is less than d_{ij} then we must assume that we have a spring which contributes the same amount of potential energy when it is compressed to a certain extent as it does when it is extended by the same amount. In this context, we are therefore obliged to replace the familiar spiral spring of previous chapters by a rigid metal beam which expands under tension and contracts under compression in such a way as to contribute a quantity to the potential energy function which is proportional to the square of the amount by which it is lengthened or shortened. It is interesting to note that this optimality criterion was aptly named *stress* by Joseph Kruskal in 1959.

11.5   Concluding Remarks

Although we have restricted our discussion of multivariate statistical techniques to a set of optimality criteria based on the sum of the squared deviations, the models discussed in this chapter may readily be generalised to a wide class of fitting criteria including the sum of the absolute deviations, the maximum absolute deviation, and the median absolute deviation criteria of Chapters 5 to 7. It is to be hoped that this book will have given readers some impression of the contribution that geometrical and mechanical models can make to their understanding of statistical techniques. Further, in an era when statistical packages are increasingly based on graphical displays and the associated statistical analysis is becoming more visual, it seems appropriate to urge the reintroduction of mechanical and geometrical ideas into the classroom treatment of the subject. However, these ideas should not be regarded primarily as potential sources of computational procedures or devices, but rather as a means of broadening students' understanding of the statistical techniques they will soon have to employ in practical situations.

References

Farebrother, R. W. (1992), Geometrical foundations of a class of estimators which minimise sums of Euclidean distances and related quantities, in Y. Dodge (Ed.), *L₁-Statistical Analysis and Related Methods*, North-Holland Publishing Company, Amsterdam, 337–349.

Fisher, N. I. (1985), Spherical Medians, *Journal of the Royal Statistical Society (Series B: Methodological)* **47**: 342–348.

Hald, A. (1998), *A History of Mathematical Statistics from 1750 to 1930*, John Wiley and Sons, New York.

Jupp, P. E. and K. V. Mardia (1980), A general correlation co-efficient for directional data and related regression problems, *Biometrika* **67**: 163–173.

Mardia, K. V. and R. J. Gadsden (1977), A small circle of closest fit for spherical data and areas of vulcanism, *Applied Statistics* **26**: 238–245.

Stigler, S. M. (1986), *The History of Statistics: The Measurement of Uncertainty Before 1900*, Harvard University Press, Cambridge, Massachusetts.

List of Figures

List of Tables

Name Index

243

Subject Index